AIR, WATER AND SOIL POLLUTION SCIENCE
AND TECHNOLOGY SERIES

HEAVY METAL COMPOUNDS IN SOIL: TRANSFORMATION UPON SOIL POLLUTION AND ECOLOGICAL SIGNIFICANCE

AIR, WATER AND SOIL POLLUTION SCIENCE AND TECHNOLOGY SERIES

Trends in Air Pollution Research
James, V. Livingston (Editor)
2005. ISBN: 1-59454-326-7

Agriculture and Soil Pollution: New Research
James, V. Livingston (Editor)
2005. ISBN: 1-59454-310-0

Water Pollution: New Research
A.R. Burk (Editor)
2008. ISBN: 1-59454-393-3

Air Pollution: New Research
James, V. Livingston (Editor)
2007. ISBN: 1-59454-569-3

Air Pollution Research Advances
Corin G. Bodine (Editor)
2007. ISBN: 1-60021-806-7

Marine Pollution: New Research
Tobias N. Hofer (Editor)
2008. ISBN: 978-1-60456-242-2

Complementary Approaches for Using Ecotoxicity Data in Soil Pollution Evaluation
M. D. Fernandez and J. V. Tarazona
2008. ISBN: 978-1-60692-105-0

Complementary Approaches for Using Ecotoxicity Data in Soil Pollution Evaluation
M. D. Fernandez and J. V. Tarazona
2008. ISBN: 978-1-60876-411-2
(Online Book)

Lake Pollution Research Progress
Franko R. Miranda and Luc M. Bernard (Editors)
2008. ISBN: 978-1-60692-106-7

Lake Pollution Research Progress
Franko R. Miranda and Luc M. Bernard (Editors)
2008. ISBN: 978-1-60741-905-1
(Online Book)

River Pollution Research Progress
Mattia N. Gallo and Marco H. Ferrari (Editors)
2009. ISBN: 978-1-60456-643-7

Heavy Metal Pollution
Samuel E. Brown and William C. Welton (Editors)
2008. ISBN: 978-1-60456-899-8

Cruise Ship Pollution
Oliver G. Krenshaw (Editor)
2009. ISBN: 978-1-60692-655-0

Water Purification
Nikolaj Gertsen and Linus Sønderby (Editors)
2009. ISBN: 978-1-60741-599-2

Environmental and Regional Air Pollution
Dean Gallo and Richard Mancini (Editors)
2009. ISBN: 978-1-60692-893-6

Environmental and Regional Air Pollution
Dean Gallo and Richard Mancini (Editors)
2009. ISBN: 978-1-60876-553-9
(Online Book)

Industrial Pollution including Oil Spills
Harry Newbury and William De Lorne (Editors)
2009. ISBN: 978-1-60456-917-9

Traffic Related Air Pollution and Internal Combustion Engines
Sergey Demidov and Jacques Bonnet (Editors)
2009. ISBN: 978-1-60741-145-1

Sludge: Types, Treatment Processes and Disposal
Richard E. Baily (Editor)
2009. ISBN: 978-1-60741-842-9

From Soil Contamination to Land Restoration
Claudio Bini
2010. ISBN: 978-1-60876-853-0

Heavy Metal Compounds in Soil: Transformation upon Soil Pollution and Ecological Significance
Tatiana M. Minkina, Galina V. Motusova, Olga G. Nazarenko and Saglara S. Mandzhieva
2010. ISBN: 978-1-60876-466-2

AIR, WATER AND SOIL POLLUTION SCIENCE
AND TECHNOLOGY SERIES

HEAVY METAL COMPOUNDS IN SOIL: TRANSFORMATION UPON SOIL POLLUTION AND ECOLOGICAL SIGNIFICANCE

TATIANA M. MINKINA,
GALINA V. MOTUSOVA,
OLGA G. NAZARENKO
AND SAGLARA S. MANDZHIEVA

Nova Science Publishers, Inc.
New York

Copyright © 2010 by Nova Science Publishers, Inc.

All rights reserved. No part of this book may be reproduced, stored in a retrieval system or transmitted in any form or by any means: electronic, electrostatic, magnetic, tape, mechanical photocopying, recording or otherwise without the written permission of the Publisher.

For permission to use material from this book please contact us:
Telephone 631-231-7269; Fax 631-231-8175
Web Site: http://www.novapublishers.com

NOTICE TO THE READER

The Publisher has taken reasonable care in the preparation of this book, but makes no expressed or implied warranty of any kind and assumes no responsibility for any errors or omissions. No liability is assumed for incidental or consequential damages in connection with or arising out of information contained in this book. The Publisher shall not be liable for any special, consequential, or exemplary damages resulting, in whole or in part, from the readers' use of, or reliance upon, this material.

Independent verification should be sought for any data, advice or recommendations contained in this book. In addition, no responsibility is assumed by the publisher for any injury and/or damage to persons or property arising from any methods, products, instructions, ideas or otherwise contained in this publication.

This publication is designed to provide accurate and authoritative information with regard to the subject matter covered herein. It is sold with the clear understanding that the Publisher is not engaged in rendering legal or any other professional services. If legal or any other expert assistance is required, the services of a competent person should be sought. FROM A DECLARATION OF PARTICIPANTS JOINTLY ADOPTED BY A COMMITTEE OF THE AMERICAN BAR ASSOCIATION AND A COMMITTEE OF PUBLISHERS.

LIBRARY OF CONGRESS CATALOGING-IN-PUBLICATION DATA
Heavy metal compounds in soil : transformation upon soil pollution and ecological significance / authors, Tatiana M. Minkina ... [et al.].
    p. cm.
Includes bibliographical references and index.
ISBN 978-1-60876-466-2 (hardcover)
 1. Heavy metals--Environmental aspects. 2. Soil pollution. 3. Soils--Heavy metal content. I. Minkina, Tatiana M.
 TD879.H4H423 2009
 628.5'5--dc22
                            2009038988

Published by Nova Science Publishers, Inc. ✢ *New York*

Contents

Preface		ix
Introduction		xi
Chapter 1	Diversity of Heavy Metal Compounds in Soils	1
Chapter 2	Methodology of Studying Heavy Metal Compounds in Soils	25
Chapter 3	Composition of Copper, Zinc, and Lead Compounds in Clean and Contaminated Soils of the Lower Don Basin (Southern Federal District of Russian Federation)	53
Chapter 4	Sorption and Distribution of Heavy Metals in Contaminated Soils	113
Chapter 5	Transformation of Heavy Metal Compounds During the Remediation of Contaminated Soils	125
Conclusions		149
References		151
Index		177

Preface

Concentrations (clarkes) of chemical elements in different components of the environment are indicative of the redistribution of the elements between the main spheres of the planet. Regional clarkes of chemical elements attest to the geochemical diversity of these spheres. The global ecological role of soil is that it, being a product of interaction between biotic and abiotic components of the environment, exerts its own considerable influence on them. This general statement acquires special significance under conditions of constantly growing technogenic impacts on the environment, including the hazardous chemical pollution. Furthermore, the real interaction between chemical elements in all the components of the biosphere is executed by various groups of chemical compounds specific for each of the components of the environment. In this context, the compounds of chemical elements in soils attract special attention.

INTRODUCTION

Concentrations (clarkes) of chemical elements in different components of the environment are indicative of the redistribution of the elements between the main spheres of the planet. Regional clarkes of chemical elements attest to the geochemical diversity of these spheres. The global ecological role of soil is that it, being a product of interaction between biotic and abiotic components of the environment, exerts its own considerable influence on them. This general statement acquires special significance under conditions of constantly growing technogenic impacts on the environment, including the hazardous chemical pollution.

The real interaction between chemical elements in all the components of the biosphere is executed by various groups of chemical compounds specific for each of the components of the environment. In this context, the compounds of chemical elements in soils attract special attention.

The composition and mechanisms of transformation of metal compounds in soils have been studied for more than 50 years. At present, these data are of great importance for adequate assessment of the modern status of polluted soils, prediction of their further changes, and development of soil remediation strategies.

The methods for studying metal compounds in soils are being refined. The first methods were based on sequential extraction procedure ensuring the extraction of metal compounds bound by the soil particles with different strengths. Though these methods were suggested long ago, they are still in use and will be used in future for assessing the strength of bonds between chemical elements and the soil particles. The strength of these bonds predetermines availability of chemical compounds for plants, their migration in the landscape, and their impact on living organisms, which is of great ecological significance.

Since the 1960s, the most widely used methods of sequential extraction of metal compounds have been directed towards the separation of two groups of metal compounds. The first group is represented by metal compounds strongly bound with soil components represented by organic substances, nonsilicate mineral compounds, and silicate minerals. The second group includes metal compounds that are weakly bound with the same soil components. These groups are extracted with the help of solvents making it possible to judge the strength of bonds between metal compounds and soil components.

Along with fractionation of metal compounds into these two large groups, the contents of mobile metal compounds are determined. Data on the mobile metal compounds are of great practical importance, as they are considered to be available for plants. Acid-extractable and exchangeable metal compounds are usually referred to as mobile metal compounds. Their concentrations in soils are indicative of the degree of soil contamination. The group of metal compounds weakly bound with organic substances may be considered a transitional group. The meaning of mobile metal compounds with organic substances is that they are active in relatively loose binding of metals of the technogenic origin; in the course of further transformations, these compounds replenish the pool of strongly bound metal compounds.

Both sequential and parallel extraction schemes characterize the same system of metal compounds in soil. The interrelationship between the results obtained by the methods of sequential and parallel extraction is quite obvious; however, it is rarely been analyzed specially. We argue that it is feasible to use and interpret the results obtained by these two methods together. On the one hand, this will make it possible to identify more exactly metal compounds extracted by individual extractants. Knowledge about the mechanisms ensuring accumulation of the particular groups of extracted metal compounds is crucial for predicting their further behavior in soils. On the other hand, the combined use of both extraction schemes offers possibility to calculate the contents of those metal compounds that cannot be separately extracted by either of these methods and, at the same time, are included in the extracted groups of metal compounds.

On this basis, a combined scheme of fractionation of soil metal compounds with the use of sequential and parallel extractions with routine extractants has been developed. The use of this scheme makes it possible to calculate the contents of some metal compounds that cannot be separately extracted.

A drawback of extraction methods for fractioning metal compounds is that the extractants used for extraction are not sufficiently selective; it is impossible to extract and identify the particular compounds. As a rule, the extractants dissolve a group of element compounds having some common property.

In the recent decades, new instrumental methods (EPR spectrometry; NMR spectroscopy; X-ray absorption spectroscopy, including the extended X-ray absorption fine structure (EXAFS) technique, etc.) have been developed; they make it possible to obtain direct information on the kind of bonds between soil components and metals. It is feasible to combine these new methods with traditional chemical extraction methods. The efficiency of such a combination is ensured by reliable qualitative data on the nature of bonds between metal compound and organic and mineral soil components and by a wealth of quantitative information on the contents of metal compounds in different soils. It is also important to study the state of metal compounds in polluted and nonpolluted soils of different nature zones on the basis of traditional extraction methods and their recent modifications combined with experimental studies of the state of plants grown on these soils.

Technogenic pressure on agroecosystems results in the accumulation of heavy metals in soils; heavy metals are redistributed between the solid and liquid soil phases. Assessments of the ecological state of polluted soils should take into account changes in the metal mobility rather than changes in their bulk contents in the soils.

The strength of metal binding in heterogeneous soil systems dictates the intensity of migration and accumulation of metals in soils. The impact of metal-contaminated soils on the entire ecosystem directly depends on the group composition of metal compounds.

Mobile forms of metals in the soil solid phase are in the state of equilibrium with metals in the soil liquid phase. This equilibrium in heterogeneous soil systems has a dynamic character, which creates conditions for further transformation of metal compounds. The transformation of metal compounds in contaminated soils depends on the chemical nature of particular metals, the degree of pollution with them, the presence of other accompanying pollutants, the soil properties, and the time factor.

The more heavy metals can be strongly fixed in the soil solid phase, the more actively they are removed from the soil solution; their availability for plants decreases. Soil serves as a barrier to the migration of metals into neighboring media.

Extensive data on the contents of different groups of heavy metals in soils have been accumulated. These data can be used to find some general regularities of the behavior of heavy metals in soils and to pose new problems for the research. Knowledge of the mechanisms of firm fixation of metals by the soil components is of particular importance, especially in the context of remediation of polluted soils. Indices of the barrier function of soils should be properly characterized. For this purpose, it is necessary to know mechanisms of the transformation of heavy metals compounds of the anthropogenic origin in soils, the mobility of these compounds, and their availability for plants. It is particularly important to have a common methodological basis for comparing data obtained by different methods and for assessing their information value. The list of heavy metal compounds arranged with respect to their availability for plants should be developed. The key problem of the particular mechanisms ensuring metal binding in soils and determining metal availability for plants has yet to be solved. In order to assess the migration capacity of heavy metals, experimental data on their concentrations in the soil–plant system are necessary. The problem of permissible concentrations of heavy metals in soils is far from being solved, and the efficient methods of remediation of highly fertile soils, including chernozems, contaminated by heavy metals have to be developed.

The problem is further complicated by the need to take into account regional variability in the contents of the particular forms of metal compounds in soils. Soils of the Lower Don basin attract special attention, as this region is one of the major agricultural producers and, at the same time, a huge industrial center exerting strong anthropogenic pressure on the environment and soils. The Lower Don region belongs to the steppe and dry steppe zones with diverse environmental conditions; the high population density; and the high concentration of industry, including metallurgical plants and coal-mining, and ore-mining factories. These industries, as well as a huge coal-fired thermal power station in Novocherkassk, are active sources of the environmental pollution by heavy metals. The steppe soils are represented by chernozems, the most fertile and valuable soils of Russia. The transformation of heavy metal compounds in these soils has its own specificity conditioned by the presence of carbonates.

This monograph is a result of experimental and theoretical studies. The materials presented in it expand the existing notions on heavy metal compounds in soils. The concepts of the groups and fractional composition of metal compounds in soils are formulated in it. The methodology for studying the group composition of heavy metal compounds in soils us suggested, and its practical application for assessing the mobility of heavy metals in the ecosystem and predicting their further behavior is discussed. An original combined scheme of fractionation of heavy metal compounds in soils is described. The general regularities and regional specificity of the behavior of heavy metals in soils are considered. A polyfunctional nature of soil components and their participation in the strong and weak fixation of metals are shown. The particular mechanisms of the transformation of metal compounds in soils in dependence on the soil properties are discussed. Soils of the Lower Don region are arranged into a genetic sequence with respect to their capacities for metal binding. The impact of soil contamination by heavy metals on the ecological status of soils and the problems of soil tolerance toward the contaminants are discussed. Approaches for the selection of appropriate ameliorants of contaminated soils are suggested on the basis of the analysis of the mechanisms of metal binding.

The theoretical considerations and methodological approaches discussed in the book may be applied in the practice of soil-ecological monitoring and assessment of the state of heavy metals in soils and plants. The may also serve as the basis for predicting and regulating the quality of contaminated soils and the crops grown on them.

Chapter 1

DIVERSITY OF HEAVY METAL COMPOUNDS IN SOILS

1.1. METAL COMPOUNDS IN SOILS AND MECHANISMS OF THEIR SORPTION

The environmental significance of chemical elements is determined by the contents of their particular compounds in soils. As early as in 1921, V.I. Vernadsky was the first to draw attention to the importance of studying the forms of chemical elements in the Earth's crust; he substantiated it in more detail in *Essays on Geochemistry* in 1934. Vernadsky wrote that the study of forms of chemical elements is the major objective of geochemistry: "The study of geochemical problems can be reduced to the study of the history of each chemical element and the mutual relationships of elements in equilibrium groups, because the permanent transition of elements from one equilibrium group into another during the historical time is a characteristic feature of the terrestrial history of chemical elements." These statements were based on logical reasoning, because the acquisition of experimental data was just started at that time. Vernadsky wrote: "The forms of occurrence were separated empirically, each of them having a specific state of atoms. In essence, these are the domains with different states of atomic systems. The state of dispersed elements is a form of element occurrence that is often ignored. However, this form plays an important role in the migration of elements. We should successively study the behavior of each chemical element in all the forms, and pay special attention to the migration of elements from one form into another." The term "forms of occurrence" or "forms of elements" signifies that a certain set of chemical elements differing from other sets in some features is considered rather than individual chemical elements themselves.

The term "heavy metal fractions" is frequently used as a synonym to the term "heavy metal forms." It implies heavy metal compounds extracted from the soil by specific chemical reagents. The methods of the particular heavy metal compounds from a soil are called fractionation methods. In the choice of reagents for extracting heavy metal compounds from the soils of natural and technogenic landscapes, the following factors should be considered: (1) the multicomponent and multiphase structure of soil; (2) the presence of the same elements in the natural soils and technogenic emissions; (3) the possibility of simultaneous reactions of heavy metals with different components of soils and technogenic emissions; and (4) the variability of the impact, intensity, and material composition of technogenic emissions in time, as well as of the spatial distribution of pollutants (Ladonin, Plyaskina, 2003).

A.P. Vinogradov (1957), V.V. Kovalskii (1961, 1970), and N.G. Zyrin (1958, 1983) showed that the behavior of metals in soils, as well as their availability to and their effect on living organisms, depends of their forms in the soil.

Heavy metals occur in soils as free cations or various chemical and physicochemical compounds (Retstse, Krystya, 1986; Berti, Jacobs, 1996). The presence of particular forms of metals in soils depends on many factors: the soil-forming material; the presence and nature of technogenic pollution sources; and the natural environmental conditions, including the pH, the redox potential, the cationic and anionic compositions of the soil solution, and the composition and properties of soil solid phases (Gray, McLaren, 2006; Kirkham, 2006).

Hydrolysis, hydrolytic polymerization (formation of polynuclear hydroxo complexes), and association with different ligands from the soil solution are the most significant processes determining the distribution of heavy metal forms in the solution.

Heavy metals in primary minerals mediate between the soil and parent rocks. In spite of the possible formation of minerals and insoluble Zn, Cu, and Pb compounds, their major forms in clean soils are in some way bound to the mineral and organic components (Adriano, 2001).

The metals considered readily form complexes with different ligands, especially with organic anions because of the structure of their electron shells. These complexes are among the most important forms of elements in soil solutions. Most complexes with organic anions are stable chelates. The complexes formed by humus acids with metal salts are relatively well soluble under neutral, slightly acid, and slightly alkaline conditions. Therefore, organometallic complexes can migrate in soils (Karpukhin, Sychev, 2005; Adriano, 2001).

Different methods are used for assessing the role of separate components of the soil solid phase and factors of the system status in binding metals to the soil. Numerous works deal with the study of relationships between sorbed metals and different soil components or other physicochemical factors: pH, specific surface, particle-size distribution, cation exchange capacity (CEC), etc. (Karnaukhov et al., 1989; Pinsky, 1992; Andersen at al., 2004; Dumat et al., 2000; Meers at al., 2006). A correlation was found of the content of Zn, Cu, and Pb in the soil (or their maximum adsorption) with the specific surface, clay, organic matter, CEC, pH, carbonates, Fe and Mn oxides, labile clay minerals, and exchangeable Ca (Machado, Pavan, 1987; Nielsen, 1990; Yin, You, Allen, 1999; Chaignon et al., 2001).

Along with the search of correlations for analyzing the effect of different factors on the content of metals in soil solid phases, methods based on model experiments with separated soil components (e.g., clay fraction, humic and fulvic acids) or synthetic analogues of soil materials or organic compounds are used (Zyrin at al., 1986; Bibak, 1994).

The interaction of different soil components with metals results in the formation of various metal compounds (Emmirich et al., 1982; Gorbatov, 1983; Retstse, Krystya, 1986; Sadovnikova, 1997; Vodyanitskii, Dobrovol'skii, 1998; Zhideeva et al., 2002; Roberts, Scheinost, Sparks, 2002; Pierzynski et al., 2005).

Metal forms are subdivided not only by the association with some soil components (organic matter, Fe and Mn hydroxides, carbonates, sulfides), but also by the binding nature with soil particles (exchangeable, specifically and chemically sorbed, occluded) and the ability to release under changing environmental factors: Eh, pH, and solution concentration. Mobile compounds (source and the nearest reserve of metals for plants), fixed compounds (potential reserve), and isomorphic impurities in minerals (strategic reserve) are also distinguished (Zyrin, 1968). In addition, heavy metal forms bound to different particle-size fractions of soils are differentiated (Motuzova, 1972). Water-soluble, exchangeable, specifically sorbed, and acid-soluble forms are not bound to a specific soil component.

The presence of the following heavy metal forms in the soil was proved (Sadovnikova, 1997):

- Soluble forms: free ions and soluble complexes of heavy metals with inorganic anions or organic ligands of different stability;
- Exchangeable forms: heavy metals mainly retained by electrostatic forces on clay and other minerals, organic substances, and amorphous compounds with the low pH of zero charge;
- Specifically sorbed forms: heavy metals mainly retained by covalent and coordination bonds;
- Forms strongly fixed by organic matter: heavy metals retained by complexation and chelation reactions on pure organic matter or that bound to Fe, Al, and Ca cations, Fe and Al oxides and hydroxides, or clay minerals;
- Heavy metals retained by Fe, Al, and Mn oxides and hydroxides as occluded cations within amorphous compounds or adsorbed on their surface; and
- Precipitates of heavy metal salts (carbonates, sulfides, phosphates) or hydroxides occurring as mixed crystals or mixtures of crystals of different elements.

The systemic analysis of the status of chemical elements in the soil reveals interrelation between their compounds. The material composition of any soil is characterized by the elementary system of chemical compounds. This is a system of compounds of a chemical element in the solid, liquid, and gaseous phases of the soil mutually related by the transformation and redistribution of matter and energy occurring at the material-phase, soil-profile, and landscape-geochemical levels (Motuzova, 1999). This system includes strongly bound mineral, organic, and organomineral compounds and mobile compounds of solid phases, soil solution substances, soil air, and biota.

The material carrier of the elementary systems of all chemical elements is the minimum volume of soil material (morphological element, horizon) containing all components of the system. The list of these components is necessary and sufficient to characterize the soil as a specific natural body (Fig. 1.1).

Mobile metal compounds are of highest ecological importance. The soil solution includes mobile substances of the liquid phase; potentially mobile compounds consist of mobile compounds of soil solid phases.

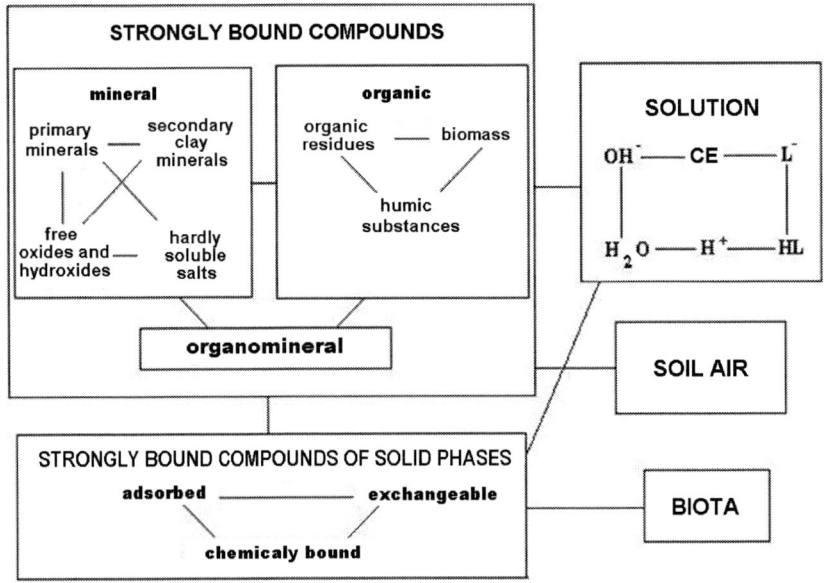

Figure 1.1. Elementary system of compounds of a chemical element in soils.

The main mechanisms of metal sorption by soils are as follows (McBride, 1981):

- The formation of insoluble metal compounds;
- The adsorption by components of the soil solid phase;
- Ion-exchange sorption;
- The formation of inner- and outer-sphere metal complexes;
- The isomorphic substitution and occlusion by Al, Si, Mn, and Fe oxides and carbonates; and
- The accumulation by living organisms.

The role of separate mechanisms in the sorption of heavy metals by soils depends on several factors: the nature and concentration of metal in the soil, acid–base and redox conditions, the content and quality of organic substances, the mineralogy and texture of soils, the activity of microorganisms, etc.

The interaction of heavy metals with soil solid phases results in the formation of new phases as precipitates of low-soluble compounds with different compositions and exchangeably and nonexchangeably sorbed forms, the retention strength of which depends on the properties of metal and soil exchange complex (SEC) (Pinsky, 1997).

Complex and diverse interactions of metals with the surface of soil solid phase are due to the wide range of surface functional groups. Different classifications are used for the mechanisms of metal fixation by soil particles. Three main mechanisms are separated for the interaction of metals with the surface of soil particles (Sposito, 1984, 1989):

1. Coulomb interaction between metal ions (hydrated cations $Me^{n+} \cdot mH_2O$, hydroxo and other complexes) with the charged surface. The metal is completely dissociated and retained in the surface diffuse layer by electrostatic attraction forces, but retains mobility and ability to move in the near-surface solution volume.
2. The formation of localized outer-sphere complexes of hydrated metal ions with surface functional groups.
   Metals bound to the surface by the first two ways belong to nonspecifically sorbed (exchangeable) cations, which can be displaced by other cations.
3. The specific sorption of metals with the formation of strong covalent bonds with the functional groups of organic and inorganic components of soils (inner-sphere complexes). Metal cations occur in the dense part of the double electric layer. Such adsorption can occur even on a likely charged surface, in spite of the generated Coulomb repulsion.

An EXAFS study identified five mechanisms of metal fixation by the soil (Brown et al., 1999; Ford et al., 2001; Manceau et al., 2000, 2002). They are as follows:

1. The formation of outer-sphere surface complexes. The sorbed ion with its inherent shell is fixed on the charged surface among diffuse ions by electrostatic forces. Sorbate particles are isolated from the metal by two layers of oxygen atoms.
2. The formation of inner-sphere surface complexes. Sorbed ions are separately bound to the sorbent surface with the participation of one or several ligands, most frequently through an oxygen atom. During the crystal growth, the sorbed ion can gradually penetrate into the sorbent structure. This sorption mechanism is implemented due to the presence of structural defects on the solid surface.

3   The formation of polynuclear surface complexes. The sorbent acts as a structural support for the newly formed precipitate. The presence of the mineral surface decreases the supersaturation level necessary for precipitation due to the similar sizes of two lattices (McBride, 1989). The sorbate cations are then polymerized on the substrate.
4   Homogeneous precipitation. The process occurs when dissolved cations are polymerized and precipitated from the solution without formation of structural bond with the sorbent. Later on, homogeneous precipitates cover the substrate as a shell (Dahn et al., 2002).
5   Diffusion into the sorbent lattice. The sorbed ion diffuses into the sorbent lattice, fills vacancies, and substitutes sorbent atoms (Sparks, 2003).

Let us consider the diversity of heavy metal forms in soils with three elements (copper, zinc, and lead) as an example.

1.2. COPPER, ZINC, AND LEAD COMPOUNDS IN SOIL SOLUTION

In a wide spectrum of pollutants, copper, zinc, and lead deserve special attention. Copper, zinc, and lead compounds are the most common environmental pollutants. These are elements of the first and second classes of ecological hazard. However, they differ in biological significance. Copper and zinc are microelements essential for living organisms; the role of lead is almost unstudied, and the need of plants for this element is not determined.

The concentration of Cu in soil solutions is 10^{-7}–10^{-6} M (Hodgson et al., 1966). In most soils, the activity of Cu^{2+} ions in the soil solution is apparently controlled by the sorption of the ion on the surface of soil particles. Complexes with organic ligands are the main forms of copper in the soil solution. They can make up to 99% of total copper in the soil solution (Hodgson et al., 1966). The activity of Cu^{2+} ions in the equilibrium soil solution is lower than that of Zn^{2+} ions by thousand times.

The complexes of Cu^{2+} with inorganic ligands play a secondary role in the formation of copper pool in the solution. The general concentration of inorganic Cu forms in the soil solution [Cu_{inorg}] corresponds to the following equation (Lindsay, 1979):

$$[Cu_{inorg}] = [Cu^{2+}] + [CuSO_4^0] + [CuOH^+] + [Cu(OH)_2^0] + [CuCO_3^0]$$

Other complexes capable to make up more than 1% of the total concentration of inorganic Cu forms include ions $CuHCO_3^+$ formed at a high pCO_2 level and a neutral reaction, $CuHPO_4^0$ at a high phosphate concentration and a neutral reaction, and $CuCO_3^0$ at a high pCO_2 level and a high pH.

Lead is one of the least mobile heavy metals characterized by a relatively low content in natural soil solutions (Kabata-Pendias, Pendias, 1989). The mobility of lead is most strongly affected by the formation of soluble chelate complexes with organic compounds and, in arid regions, chloride ions (Adriano, 2001).

According to Schulthess and Huang (cited from Adriano, 2001), Pb^{2+} ions are predominant inorganic lead ions in solutions with low pH. The $PbOH^+$ ions prevail at pH 7.5–9.5, $Pb(OH)_2$ at pH 9.5–11.0, and $Pb(OH)_3^-$ under more alkaline conditions.

According to Hodgson et al. (1966), the concentration of Zn in the soil solution is 10^{-6}–10^{-4} M. In lysimetric waters from soils of natural and technogenic landscapes, Zn compounds bound to suspended particles prevail over those in the true solution (Arzhanova et al., 1981).

The content of Zn in natural waters and soil solutions can be controlled by the solubility of phosphates (Vorob'eva, Rudakova, 1981). According to Lindsay (1979), the total concentration of inorganic Zn compounds in solution can be described as follows:

$$[Zn_{inorg}] = [Zn^{2+}] + [ZnSO_4^0] + [ZnOH^+] + [Zn(OH)_2^0] + [ZnHPO_4^0]$$

In water extracts, zinc can also be present in the anionic form (Camerlinc, Kiekens, 1982), but the zincate ions ($HZnO_2^-$ and ZnO_2^{2-}) play no significant role in most soils (Lindsay, 1979).

The ratio of Zn ions in solution depends on the pH level and the phase solution composition: Zn^{2+} ions are predominant inorganic Zn forms in acid and neutral soils, and Zn hydroxo complexes prevail under more alkaline conditions (Prokhorov, Gromova, 1971; Loganathan et al., 1977). Along with hydroxo complexes, the soil solution can contain zinc complexes with inorganic ligands: chlorides, phosphates, nitrates, and sulfates.

Complexes of Zn with soluble soil organic substances significantly contribute (75% and more), as was noted by Gapon as early as 1937. The degree of complexation of zinc, as well as other metal ions, correlates with the concentration of soluble organic matter in the liquid phase (Lindsay, Norvell,

1972; Shuman, 1983). A higher capacity of binding with the low molecular weight fraction of water-soluble organic substances than with the higher molecular weight fractions is frequently observed for Zn^{2+} ions (Arzhanova et al., 1981).

The formation of stable complexes in the soil liquid phase is typical for the Zn^{2+}, Cu^{2+}, and Pb^{2+} ions; therefore, their mobility and availability to plants and other living organisms under natural conditions depend on soluble organic and inorganic ligands. The presence of organic acids increases the solubility of heavy metals, including due to the decrease in the pH level. Metals can migrate down the profile predominantly as organometallic complexes. This is especially typical for podzolic soils. Depending on the soil and its reaction, 76–99% of Cu, 84–99% of Mn, and 5–90% of Zn occur in the soil solution as complexes with organic compounds (Karpukhin et al., 1980; Kaurichev et al., 1983; Shestakov et al., 1984; Shestakov et al., 1989). The share of metal complexes is higher in the soils with the high content of organic matter and higher pH levels. Complexation is more manifested in the upper soil horizons; the concentration of metals in the soil solutions of these soil horizons is also higher than in the lower horizons.

1.3. MOBILE COPPER, ZINC, AND LEAD COMPOUNDS AND THEIR AVAILABILITY TO PLANTS

Mobile compounds of chemical elements form an important group of chemical substances in the soil. They enable the soil to perform its main ecological functions as a naturalistic body and a source of fertility and protection from contamination.

The term "mobility" was applied by N.M. Sibirtsev for chemical elements in the soil as early as the 19th century; however, it is still used ambiguously. In geology and geochemistry, it implies the capacity of an element to participate in migration with water or other flows in solution or in the sorbed state in the solid phase. In agricultural chemistry and soil science, the terms "labile" and "mobile" are synonyms and frequently equivalent to the notion "available to plants". Therefore, mobile compounds of chemical elements are extracted from soils by reagents to some extent simulating the effect of natural waters and plant root exudates. The selection of extractants for mobile metal forms depends on the soil type and metal properties. Ya.V. Peive and G.Ya. Rin'kis (1975) recommended the use of separate solutions for extracting the mobile forms of each microelement (e.g., 0.1 N H_2SO_4 for Mn, 1 N HCl for Cu, 1 N KCl for Zn, etc.), which were empirically selected so that the content of the extracted element was most similar to its input to plants. 1 N CH_3COONH_4 solutions with different pH have found a wide use for the solubilization of mobile microelement compounds: pH 4 (Baron, 1955), pH 3.5 (Arinushkina, 1961), and pH 4.8 (Krupskii, Aleksandrova, 1964). In different sequential fractionation schemes, mobile microelement compounds are extracted first.

In methodological guidelines (Methodological Guidelines..., 1992), it is recommended to determine the mobile forms of metals (Cu, Zn, Ni, Co, Pb, Cd) using a 1 N ammonium acetate buffer solution (AAB) with pH 4.8, 1 M HCl, and 1 M HNO_3. These reagents are used by the Agrochemical Service of RF for extracting plant-available microelements and assessing the soil supply with these elements.

It is believed that the extraction of mobile metal compounds from soils by different reagents is comparable with the effect of natural waters and plants. At the same time, D.N. Pryanishnikov (1953) noted that roots operate on the soil for a longer time period than a solvent and remove reaction products from the impact zone, while a solvent operates for a short time and its dissolution products remain in the solution. Consequently, there is no universal solvent that would allow predicting the content of a specific element sorbed by all plants. All solvents recommended for the extraction of the mobile compounds of chemical elements are defined by Pryanishnikov as "conventional" and can be used only for comparing with the actual supply of plants with these elements. Thus, it is hardly possible to find an extractant adequately characterizing the input of a chemical element from the soil into plants. The development of a universal extractant for a large group of potentially hazardous chemical elements is even more problematic (Il'in, 2006).

The availability of soil substances to plants is determined by their reserve, concentration in the soil solution and rate of desorption from the surface of soil particles (Ovcharenko, 1996; Kuz'mich et al., 2000). However, model calculations with account for these factors are complicated and ambiguous. Although some authors (Frid, 1996) believe that the methods of extracting mobile nutrients (including microelements) from soils with neutral salt solutions, ammonium acetate, EDTA, DTPA, and diluted acids are unsatisfactory, the extraction with these reagents is used in many countries as a standard method of determining the content of chemical elements available to plants in soils.

The first extractants for the mobile forms of heavy metals were found empirically on the basis of correlations between the extracted element compounds and their contents in plants. However, the selection of extractants for mobile forms of microelements for the ecological prediction of pollution consequences should be based not only on agrochemical, but also on soil-chemical principles.

The direct analysis of acid extracts has found wide use for acquiring prompt information on the ecological status of soils not only in technogenic, but also in natural landscapes. However, the concepts of solubility of natural microelement compounds in these extracts are not finally developed (Motuzova, Degtyareva, 1991; Protasova, Gorbunova, 2006). Il'in (2000) believes that the hydrochloric acid solution extracts the so-called near pool of heavy metals from the soil, which can feed the migration of pollutants to adjacent environments through trophic chains. Rin'kis (1962) attached great importance to the latter factor. Therefore, it was logical to include this extractant in soil-agrochemical studies. Scientists of the COMECON countries developed draft the maximum concentration limits (MCLs) for mobile metal forms (in a 1 N HCl extract) (Chudzhiyan et al., 1988), which remained unfinished. However, the draft document found use in the ecological practice. The fact that the dissolving capacity of 1 N HCl is similar to that of acid rains was also an important argument for its use, ant it was supposed that the pool of potentially available heavy metals in the soil could be readily determined using this extractant.

The relative content of the mobile forms of most microelements is usually estimated at tenths of a percent or few percents and rarely reaches 10%. The content of mobile lead in ordinary chernozem is 2.5% (Samokhin, 2003); a similar content is typical for gray forest soils in Tula oblast (Perelomov, Pinsky, 2003). The share of mobile zinc is 5–8% of the total element content in brown and soddy soils of the Far East region of Russia (Motuzova, Smirnova, 1983), 7–8% in soddy-podzolic soils of Moscow oblast, 1.4% in typical chernozem of Kursk oblast, 0.8% in ordinary chernozem of Chelyabinsk oblast, 1.6% in dark chestnut soil of Kherson oblast, 2.2% in krasnozem of Georgia, 3.6% in sierozem of Samarkand oblast (Rerikh, 1976), and 0.1–1% in red ferrallitic soils of the Southern Viet Nam and Cambodia (Van, 1995).

Mobile metal compounds in different soils are given in Table 1.1. The data show that the relative contents of mobile Cu, Zn, and Pb forms in different soils mainly vary from tenths of a percent to 14%.

Table 1.1. Mobile compounds of Cu, Zn, and Pb in soils, % of the total content

Soil, region	Element	Exchangeable	Acid-soluble	Source
Sandy loamy soil, Komi Republic	Cu	2.7	21.3	El'kina et al., 2001; 2008
	Pb	7.9	37.2	
Loamy soddy-podzolic soil, Belarus	Cu	3.3	64	Golovatyi, 2002
	Pb	5.4	75.2	
	Zn	9.3	70	
Soddy-podzolic soil, Moscow oblast	Pb	2.7	37.3	Nosovskaya et al., 2001
Calcareous alluvial soddy soil, Tomsk oblast	Cu	3.8	50.5	Izerskaya, Vorob'eva, 2000
	Zn	2.1	13.8	
Gray forest soil, Ryazan oblast	Cu	0.3	14.6	Kuznetsov et al., 1995
	Pb	1.6	39.4	
	Zn	2.6	14.4	
Light gray forest soil, Nizhni Novgorod oblast	Pb	4.3	50.4	Dabakhov et al., 1998
Leached chernozem, Central Chernozemic Zone	Pb	6	8	Protasova, Gorbunova, 2006
Typical chernozem, Central Chernozemic Zone	Pb	10	15	
Ordinary chernozem, Central Chernozemic Zone	Pb	11	20	
Clay loamy typical chernozem, Belgorod oblast	Cu	1.9	30.6	Pendyurin, 1986
	Pb	3.3	37.1	
	Zn	0.6	16.0	
Meadow-chernozemic soil, Kursk oblast	Zn	14	64	Pollutants ..., 1988
Clay loamy ordinary chernozem, Rostov oblast	Cu	0.8	12.0	Khoroshkin, 1979
	Zn	0.6	8.8	
Clay loamy chestnut soil, Rostov oblast	Cu	1.4	11.3	
	Zn	0.9	9.4	
Clay loamy southern chernozem, Rostov oblast	Cu	0.5	17.6	Nikityuk, 1998
	Pb	6.8	12.0	
	Zn	0.2	12.6	

Low concentrations of exchangeable forms are usually typical for Cu; higher concentrations are observed for Pb (Table 1.1). A higher content of acid-soluble forms is also noted for Pb.

1.4. METAL COMPOUNDS BOUND TO ORGANIC MATTER

Soil organic matter plays an important role in the binding of metals, which results in a high correlation between the contents of metals and organic substances in the soil (Schizer, Skinner, 1967; Steinnes, Njastad, 1995; Karpukhin, 1998). The interaction of heavy metals with humus can occur through ion exchange, complexation, and adsorption. One kilogram of humic acid separated from peat can adsorb 40–200 g metals. Fulvic acids have an even higher adsorption capacity (Aleksandrova et al., 1970; Aleksandrova, Naidenova, 1970; Pickering, 1980; Donisa, Steinnes, Mocanu, 2001; Donisa, Mocanu, Steinnes, 2003; Huang et al., 2005; Violante et al., 2002).

L.N. Aleksandrova (1980) separated three types of possible organomineral compounds in the soil. One of them—heteropolar salts—includes humates and fulvates resulting from exchange reactions occurring during the interaction of humus acids with metal cations. An ionic bond is formed between the metal and organic matter. Carboxyl and phenolic groups are most active in the fixation of metals. The resulting compounds are low stable, and adsorbed metals can readily leave them through ion exchange reactions (Motuzova, 1988).

The second type of organomineral derivatives is represented by complex salts. In this case, heavy metal cations displace hydrogen ions of some functional groups, partially enter into the inner sphere of the complex salt, and cannot participate in exchange when occur in the anionic part of the molecule. The stability of heavy metal complexes with humic and fulvic acids usually increases with increasing pH. The adsorption strength of Pb by humic acids increases with increasing degree of humification.

The third type of organomineral compounds of heavy metals includes adsorption and chemisorption complexes on the surface of solid particles. Clay minerals and Fe and Mn hydroxides actively participate in their formation. Adsorption complexes are very stable and can be decomposed only under strongly acid or strongly alkaline conditions.

The complexing capacity of Zn, Cu, and Pb and their relative stability are due to the structure of their electron shells. These elements belong to the transition metal group, have a relatively high electronegativity, and readily form covalent bonds.

The binding of metal ions by organic matter results in the release of a proton from untitrated functional groups (Stevenson, 1976), which do not enter into cation exchange with alkali and alkali-earth cations.

Two mechanisms of metal sorption by soil humic acids are considered (Stevenson, 1976; Van Dijk, 1971). At low pH, nonhydrolyzed metal ions displace protons from functional groups of humic acids; at higher pH values, protons are released from water molecules with the formation of metal hydroxo complex with humic acid.

According to Aleksandrova (1980), the metal displaces a proton from an acid functional group in the inner anionic part of the molecule. Other metal ions are bound to free functional groups and can participate in cation exchange reactions. The resulting metal complexes with organic matter are called heteropolar organomineral complexes. The ratio between the metal atoms capable of exchange and those bound into stable complexes depends on the metal nature, the ionic strength of solution, and the pH (Stevenson, 1976).

It was found that the interaction of metals with soil organic matter occurs via the formation of stable metal complexes with COOH, OH, and NH_2 groups (Manskaya et al., 1958, Stevenson, 1976; Leenheer, 1999; Tarr et al., 2000; Bilali, Rasmussen, Fortin, 2001; Chakrabarti, 2001; Choi et al., 2008). Chelate structures can be formed in this case, including the interaction of metal with two COOH groups or COOH and OH groups, analogously to complexes with phthalic and salicylic acids or amino groups (Manceau et al., 2000; Wong at al., 2007; Amit, Chen, 2008).

The formation of chelates is confirmed by IR spectroscopic data. The study of EPR spectra showed that Cu also forms porphyrin-type complexes with humic and fulvic acids (Bartashevsky et al., 1971; Vincler et al., 1971; Cheshire et al., 1977). For Zn, the formation of coordination covalent bonds with OH groups and an electrovalent bond with a COO group was revealed (Tan et al., 1971). Metal ions can bind separate organic molecules as intermolecular bridges to form polymeric structures (Zunio et al., 1975). The formation of biligand (bidentate) Cu complexes with humic acids (HAs) of the type HA–Cu–HA was considered by V.V. Demin (1994). Hering and Morel (1988) concluded that Cu and Ca in most cases form complexes with different functional groups of humic acids. Manceau, Marcus, and Tamura (2002) showed that lead could form bidentate complexes by chelating with functional groups of aromatic rings.

The stability of metal complexes with humic acids depends on the ion nature and external factors, e.g., pH and ionic strength. The estimates obtained for the stability constants of heavy metal complexes with humic and fulvic acids by different authors vary in a wide range, which can be related to problems in the separation and purification of the corresponding acids.

The microelement compounds bound to organic matter are isolated using different methods, which can be divided into two groups: (1) the oxidation of organic matter with hydrogen peroxide or nitrogen dioxide vapor and calcination at 450–500°C in the air followed by the extraction of released microelements with a reagent and (2) the extraction of soil organic compounds by alkaline solutions of sodium hydroxide, potassium pyrophosphate, and EDTA in a sodium hydroxide solution and determination of microelements in the extract.

1.5. METAL COMPOUNDS BOUND TO FE, AL, AND MN (HYDR)OXIDES

Fe, Mn, and Al oxides and hydroxides occur in the soil as individual phases or films covering clay minerals. Long-term studies showed that these hydroxides play an extremely important role in the fixation of heavy metals (Mellis, Casagrande, Cruz, 2003). Metals can be sorbed by hydroxides in amounts exceeding the CEC and even on positively charged surfaces. The adsorption capacity of freshly precipitated hydroxides is higher than that of crystalline hydroxides; the adsorption of metal hampers the crystallization of amorphous hydroxides (McBride, 1981). It should be kept in mind that crystallized particles of iron oxides and hydroxides play a major role in the strong fixation of metals in soils of different genesis (Vodyanitskii, Dobrovol'skii, 1998; Pinsky, 1992). The adsorption capacity of Fe hydroxide is higher than that of Al hydroxides. The CEC is 160 µM/g for amorphous $Fe(OH)_3$, 50 µM/g for amorphous $Al(OH)_3$, and 230 µM/g for *birnessite* (MnO_2) (Brummer et al., 1983).

Fe, Al, and Mn (hydr)oxides can retain appreciable amounts of microelements due to sorption, coprecipitation, and occlusion. Specific sorption accompanied by the displacement of protons from surface functional groups is considered to be the main mechanism of metal interaction with hydroxide surface. Adsorption occurs on the surface sites with uncompensated valence forces, whose development is frequently related to defects in the crystal lattice. The specific sorption of metals is characterized by an abrupt increase in the amount of sorbed metal in a narrow pH range. According to Davis and Leckie (1978), the sorption of copper increases from 10 to 100% in the pH range from 5 to 6. The complete adsorption of metals occurs at pH values lower than the precipitation pH of the corresponding hydroxides.

Iron hydroxides can capture other metal ions at the precipitation due to adsorption and chemisorption. The coprecipitation occurs via the following mechanism: polynuclear complexes are formed during the hydrolysis of iron ions, which are next polymerized with the precipitation of iron hydroxide. A fresh layer of polymer ions is formed on the surface of freshly precipitated hydroxide. An ESR study showed that Cu coprecipitated with Al gel enters into the aluminum–oxygen octahedrons of hydrargillite. It is presumed that Cu forms a bond with oxygen by substituting a proton in an OH group of the gel (McBride, 1981).

Copper ions sorbed by the Fe hydroxide surface can pass into the occluded state with time. The occlusion mechanism involves the capture of solid-phase microelement oxides by precipitating Fe, Al, and Mn hydroxides so that the occluded microelement compounds fall within another solid phase.

Two approaches are used for extracting this group of microelement compounds. One of them is based on the dissolution of Fe, Al, and Mn hydroxides in a strongly alkaline solution (these elements form soluble hydroxo complexes under alkaline conditions); the other approach is based on their reduction by a corresponding reagent. Two requirements are imposed on the reagent used for the extraction of microelement compounds bound to Fe, Al, and Mn (hydr)oxides: it should not dissolve the soil silicate component and should dissolve oxidized nonsilicate compounds as completely as possible. Therefore, the Mehra–Jackson method (extraction with a dithionite–citrate solution) and the Endredy photolytic method (a version of the Tamm method with the use of UV irradiation) are most widely used.

1.6. Metal Compounds in Primary and Secondary Minerals

Clay minerals are powerful accumulators of microelements in the soil. The elements are partly immobilized in the exchangeable form and partly fixed irreversibly. Clay substances can retain heavy metals due to ion exchange, chemisorption, precipitation, and occlusion (Motuzova, 1988; Ford, Scheinost, Sparks, 2001).

The capacity of clay minerals for the ion-exchange adsorption of heavy metals is due to the following properties of minerals: (1) the heterovalent isomorphic substitution of ions in silicon–oxygen tetrahedrons and aluminum-oxygen octahedrons, which creates an excess charge on the surface of clay minerals; (2) the presence of uncompensated charges in defective voids of minerals; and (3) the presence of unsaturated valences on the crystal surfaces, edges, and corners.

The sorption of metals by minerals is highly effective and is manifested as an aging effect, which usually decreases their solubility, extractability, and biological availability with time (Lavrent'eva, 2008). It was found that Pb is more selectively adsorbed on Fe oxides, halloysite, and imogolite compared to kaolinite and montmorillonite (Adriano, 2001). A higher tendency of Mn oxides toward the sorption of Pb was revealed (Kabata-Pendias, Pendias, 1989). The following series of SEC components in accordance with their effect on the sorption of metals were reported (Abd-Elfattah, Wada, 1981):

for Cu ions: iron oxides > halloysite, allophone-imogolite > humus, allophane, kaolinite > montmorillonite;

for Zn ions: iron oxides > halloysite, allophone-imogolite > kaolinite > allophone, humus > montmorillonite.

If all adsorption bonds were used only for heavy metals, 1 kg of montmorillonite would adsorb 1500–2500 mg of zinc or copper (Il'in, Stepanova, 1980). According to their ability to be sorbed by clay minerals, metals form the following decreasing series: Pb > Zn > Cu (Motuzova, 1988; Manucharov et al., 2001).

The affinity of clay minerals for cations bound to the surface depends on the charge, radius, polarizability, and hydration energy of the hydrated ion. The study of Zn sorption on three main types of clay minerals (kaolinite, illite, and montmorillonite) showed that the CEC of the minerals varies in accordance with their structural features in the series: montmorillonite > illite > kaolinite; the exchange rate varies in the series: kaolinite > montmorillonite >> illite (Pickering, 1980). Deviations from the established tendencies are related to the specific sorption of cations, steric hindrances, and the formation of complex $MeOH^+$ ions. The study of the process kinetics using a Zn^{65} label showed that cation exchange is the predominant mechanism of interaction between Zn and clays. A similar conclusion was drawn in the study of Mg and Zn exchange on bentonite and illite, Zn being more strongly bound to illite, and Mg to bentonite (Singhal, Kumar, 1977). The sorption of heavy metals on Na-montmorillonite occurred in accordance with a simple exchange model (Farrah, Pickering, 1977).

The structure of aqua complexes of metals on the surface of layered silicates was established by the EPR method (Tarasevich, Ovcharenko, 1975; Tarasevich, 1988). A mobile aqua complex of copper $[Cu(H_2O)_6]^{2+}$ can be formed on the surface of kaolinite. The partial substitution of water molecules in the complex by oxygen atoms from the octahedral and tetrahedral lattices is possible under the mechanical–chemical activation of the surface. In this case, the complex is more strongly bound to the mineral structure due to the formation of two coordination bonds. The chemisorption of Pb at pH > 7 was noted due to the interaction with the SiOH and AlOH groups in edge positions (Hildebrand, Blum, 1974).

The nonexchangeable adsorption of metal cations is related to their penetration into the montmorillonite lattice and substitution of aluminum with the formation of complex metal ions like $ZnOH^+$ or $CuOH^+$ and the following nonexchangeable sorption within or on the surface of clay structures. Free or complex ions of zinc can enter into the inner potential-determining layer of minerals through the distortions caused by aluminum–hydroxyl octahedrons. Montmorillonite can retain the element as $ZnOH^+$ complex ions in the octahedral positions unoccupied by aluminum. Because of the fixation of metals in the clay mineral lattices, their concentration in the clay fractions of soils is frequently higher than in the whole soil, and the share of microelements in the clay fractions reaches half their total content and more.

The separation and determination of this metal fraction are performed using two methods: (1) the share of microelement compounds in primary and secondary minerals can be calculated from the difference between the total content of microelements and the sum of their contents in all separated fractions (readily soluble and exchangeable compounds; those bound to organic matter; Fe, Al, and Mn (hydr)oxides; carbonates, sulfides, etc.); (2) the mineral matter remained after the separation of the above fraction can be decomposed using methods of bulk analysis, including acid digestion (necessarily with HF), fusion, etc., and the content of metals can then be determined.

1.7. METALS IN DIFFICULTLY SOLUBLE PRECIPITATES AND NATIVE MINERALS

The own minerals of microelements are usually present in soils in small amounts. For example, zincite (ZnO), covelline (CuS), and bornite (Cu_5FeS_4)

are extremely rare minerals. Increased concentrations of Ga, Sr, and Ba can be observed in plagioclases; Cr, V, Ni, Co, and Mn, in pyroxenes; Cr, V, and Cu, in spinels; etc. Most own minerals and insoluble compounds of copper (carbonate, azurite, hydroxide, malachite, tenorite, copper ferrite, and different phosphates and sulfates) are too soluble and unstable in normal soils (Lindsay, 1979). However, the metals released during the weathering of minerals from parent rocks or the transformation of compounds get in the soil because of contamination can precipitate as own minerals (Vorob'eva, Rudakova, 1981; Schlegel at al., 2001; Ford at al., 2001). Under natural conditions, galenite (PbS), cerussite ($PbCO_3$), and anglesite ($PbSO_4$) are the main forms of occurrence of lead in soils (Adriano, 2001).

According to Lindsay (1979), the own minerals of Zn (hydroxides, smithsonite, zincite, villemite, and hoplite) are unstable in most soils. Soil solutions are usually unsaturated with respect to difficultly soluble Zn precipitates (Rukhovich, 1993).

Fractionation methods show that a major part of heavy metals is released during the decomposition of carbonates and hydroxides (Tessier et al., 1979; Tessier et al., 1985). In some soils, carbonates can be the major sorbents of microelements. According to some authors (Fengxiang, Arieh, 2007), the fraction bound to carbonates is the main solid phase for many metals (Cd, Pb, Zn, Ni, and Cu) in arid and semiarid soils, especially contaminated ones. The compounds of Cd, Pb, Zn, Ni, and Cu bound to carbonates in the metal-contaminated arid soils of Israel make up 60–80%, 50–60%, 40–60%, 30–40%, and 25–36%, respectively.

In calcareous soils, metals can form compounds with carbonates due to coprecipitation and occlusion, as well as to the formation of difficultly soluble surface compounds. They can also be sorbed on oxides (predominantly iron and manganese oxides), which settle on the surface of carbonates (Sadovnikova, 1994).

The highest affinity to carbonates was observed for Cu, Pb, and Zn. The precipitation of carbonates and hydroxides can be accompanied by the coprecipitation of metals and their inclusion in the structure of the mineral formed (Zyrin et al., 1986).

Metal ions can affect the precipitation of carbonates. The ionic radius of Cu (0.73 Å) can hamper the isomorphic substitution of Ca^{2+} in $CaCO_3$, but it favors the substitution of Mg^{2+} (0.72 Å) or Fe^{2+} in $MgCO_3$ and $FeCO_3$, respectively. In alkaline soils with pH > 7.9, the solubility of zinc is controlled by the precipitation of hydroxide or carbonate (Singh, Abrol, 1985).

Some phosphate minerals contain large amounts of microelements (Hettiarachichi, 1998). A high content of lead (1–35% PbO) was found in the phosphate concentrates separated from Ferralsols (laterite podzolic soils). At high concentrations of copper and phosphate ions in soil solutions with pH > 7, copper phosphate $Cu_3(PO_4)_2$ can be stable in soils. Vorob'eva and Rudakova (1981) showed that the solubility of Zn in surface waters and soil solutions can be controlled by the phosphate $Zn_3(PO_4)_2 \cdot 4H_2O$. In Zn-deficient soils, its solubility is controlled by adsorption; in soils with increased Zn contents and relatively high concentrations of silica and phosphates in solutions, the formation of Zn_2SiO_4 and $Zn_3(PO_4)_2 \cdot 4H_2O$ is possible (Dhillon, Dhillon, 1984). On the whole, the formation of own minerals in some soils is more typical for Zn and Pb than for Cu.

In the soils developed under humid conditions, sulfides, sulfates, and chlorides are of little importance, but they can play a major role in the behavior of microelements in arid climatic zones (Minkina et al., 2003). Metal ions (mainly Fe^{2+}, Mn^{2+}, Hg^{2+}, and Cu^{2+}) form sulfides relatively stable under acid and neutral reducing conditions. Other heavy metals, like Zn, can readily coprecipitate with iron sulfides. The precipitation of metal ions as sulfides is an important mechanism of regulating the concentrations of metal ions in the soil solution.

1.8. COPPER, ZINC, AND LEAD COMPOUNDS IN DIFFERENT SOILS

The levels and proportions of microelements in soils are determined by lithogenic and pedogenic processes. Data on the distribution of metal compounds in different soils are given in Table 1.2.

On the basis of generalized experimental and literature data on the fractional composition of microelement compounds (Zyrin et al.,1979; Shuman, 1980; Reshetnikov, 1990; Shibaeva, 1990; Filatova, 1992; Ma, Uren, 1997; Adriano, 2001; Li et al., 2001; Perelomov, Pinsky, 2003), it was proposed to classify soils in accordance with the leading factors determining the element compound ratio (Motuzova, Aptikaev, 2006). Two main groups of soil are separated, with the lithogenic and pedogenic factors being the major factors affecting the development of the microelement compound systems.

Table 1.2. Cu, Zn, and Pb compounds in some soils, % of the total content (average, interval estimate)

Soil, region	Element	Exchangeable	Carbonate-bound	Bound to Fe-Mn (hydr)oxides	Bound to organic matter	In primary and secondary minerals	Source
Soddy-podzolic soil, Middle Urals	Cu	3-5	----	24-32	14-34	49-67	Reshetnikov, 1990
Sandy loamy soddy-podzolic soils, Moscow oblast	Cu	0.3	5	29	4	54	Plekhanova et al., 2001
	Pb	5	11	14	8	55	
	Zn	1	33	11	0.9	51	
Sandy clay soddy-podzolic soil, Belarus	Cu	4		41	13	42	Golovatyi, 2002
	Pb	6		7	38	50	
	Zn	5		8	60	27	
Gray forest soils, Tula oblast	Zn	-	5	6	35	54	Perelomov, Pinsky, 2003
	Pb	-	7	31	29	34	
Typical chernozem, Kursk oblast	Cu	-	-	35	0.6	59	Pampura et al., 1993
	Zn	6	0.2	3	6	91	
Dark chestnut soil, Kazakhstan	Zn	3	12	31	20	50	Panin, Kushnareva, 2007
Brown forest soils, Western Georgia	Cu	to 5	-	51-74	to	26-41	Zyrin et al., 1979
	Zn	to 2	-	25-42	to 2	64-93	
Zheltozems, Western Georgia	Cu	1-2	-	25-56	2-5	42-74	
	Zn	1-2	-	19-50	1-3%	55-84	
Krasnozems, Western Georgia	Cu	0.4-2	-	25-62	0.5-2	39-76	
	Zn	0.4-1	-	26-58	0.4-3	48-75	
Haplorthods, sandy, Poland	Pb	0.5	3.5	15	19	62	Chlopecka, 1996
Typic Udifluvent, USA	Cu	0.3	-	46	3	51	Shuman, 1980
	Zn	0.3	-	26	3	70	
Typic Hapludult, USA	Cu	0.6	-	12	14	74	
	Zn	5	-	19	11	65	
Typic Paleudult, USA	Cu	0.6	-	22	9	69	
	Zn	3	-	15	11	71	
Typic Quartzipsamment, USA	Cu	0.8	-	66	13	21	
	Zn	4.1	-	39	10	46	

Among the soils with predominant pedogenic processes, the soils with a high effect of biogenic factors on the development of the system of microelement compounds and those with a moderate effect of biogenic factors are distinguished.

The soils with the leading role of the lithogenic factor are characterized by the predominance of strongly bound element compounds in silicates. In the soils with the leading role of the pedogenic factor, element compounds bound to Fe, Al, and Mn (hydr)oxides are predominant; their content is indicative of the transformation of original minerals in the soil. The soils with increased contents of mobile microelement compounds (more than 10% of their total content) are included in the group of soils with the active manifestation of the effect of biogenic factors. If the mobile microelement compounds make up less than 10% of their total content, the effect of biogenic factors on the stat of microelements is classified as moderate.

Podzols of the Kola Peninsula developed on the sandy Scandinavian moraine present an illustrative example of soils in which the lithogenic factor had a leading effect on the formation of metal compounds. The physical weathering prevails over the chemical weathering under the hard climatic conditions of the northern regions. Their content of silica is 66–70%. The content of Fe, Al, and alkali-earth and alkali metals is relatively high, an appreciable share of them remaining in primary minerals resistant to weathering. The content of nonsilicate Fe compounds is estimated at several percents of the total element content.

The soils inherit the main properties of parent rocks. Their content of nonsilicate compounds of typomorphic elements is low: Fe and Si compounds make up less than 1%; the share of Al is 1–8%. The total content of copper, zinc, nickel, and arsenic in the regional soils is high. The major part of minerals (85–90% of their total amount) is retained by silicates.

The effect of lithogenic factors increases the content of zinc in primary and secondary minerals of Typic Udifluvent, Typic Hapludult, and Typic Paleudult (United States): 65–72% of its total content (Shuman, 1980), and in red-colored soils of Australia: 75–87% (Ma, Uren, 1997).

In chernozems and chestnut soils of the Lower Don basin, lithogenic factors also significantly affect the state of metals. The soils developed on loess-like sediments in the province of Ciscaucasian plains, where the effect of parent rocks of Caucasian geochemical provinces is manifested, are characterized by an increased content of microelements. Strongly bound metal compounds prevail in the structure of primary and secondary minerals (56–83%).

The group of soils with the leading effect of pedogenic factors includes most soils developed on well-weathered rocks (deeply reworked mantle and loess-like loams, old weathering crusts), where the share of metal compounds retained by Fe, Al, and Mn (hydr)oxides is higher than 50% of their total content. These soils include burozems, zheltozems, and krasnozems of Western Georgia and Typic Quartzipsamment (United States). In these soils, metal compounds make up 51–74%, 25–56%, 25–62%, and 66 % of the total content for copper and 25–42%, 19–50%, 26–58%, and 39% for zinc, respectively (Zyrin et al., 1979; Shuman, 1980). High relative contents of Cu, Pb, and Zn compounds bound to iron (hydr)oxides are also observed in the red-colored soils of Australia and New Zealand (Li et al., 2001).

The strong effect of biogenic factors on the formation of metal compounds in soils is manifested in two ways: (a) in an increased content of metals in soil organic matter and (b) in a relatively high content of mobile metal compounds. Metals in organic substances are a dynamic component of metal compounds in soils, because these substances are subjected to regular transformations in the soil. The accumulation of these metal compounds in soils is frequently revealed at the decelerated mineralization of organic matter. Therefore, a high share of these Cu compounds (25–40%) is observed in the tundra soils of the Novaya Zemlya (Filatova, 1992), as well as in podzols of the Upper Volga basin (1/3 to 1/2 of their total content) (Shibaev, 1990).

The relatively increased content of mobile metal compounds can be caused by their compounds weakly retained in the exchangeable form as organomineral complexes (the content of which is strongly affected by organic substances). In calcareous soils, metal compounds weakly retained by carbonates also contribute. The highest effect of these factors is manifested in polluted soils. For example, copper compounds bound to organic matter make 14–34% in the contaminated soddy-podzolic soils of the Middle Urals (Reshetnikov, 1990).

In most soils, the content of metal compounds bound to organic matter is lower than that of compounds bound to nonsilicate Fe, Al, and Mn forms by 4–7 times, which can be related to the ratio between the formation, mineralization, and transformation rates of organomineral metal compounds. An increase in the share of these compounds is observed is some soils when going from north to south.

The content of mobile metal forms in soils is usually low; it does not exceed 10% of the total element content, which is due to its strong retention by Fe, Al, and Mn (hydr)oxides and, to a lesser extent, by humus.

Thus, the ecological significance of metal compounds in the soil is due to their compounds strongly fixed in silicate minerals (largely formed under the effect of lithogenic factors) and retained by nonsilicate Fe and Mn compounds (largely formed under the effect of pedogenic factors). Pedogenic factors, primarily organic substances, have the major effect on the formation of weakly bound metal compounds in clean and especially contaminated soils.

Chapter 2

METHODOLOGY OF STUDYING HEAVY METAL COMPOUNDS IN SOILS

2.1. APPROACHES TO THE DETERMINATION OF HEAVY METAL COMPOUNDS

The assessment of the ecological status of contaminated soils is reduced to the revelation of changes in the mobility of metals rather than an increase in their total content. Metals arrived in the soil participate in simultaneous chemical interactions with components of the heterogeneous soil system. These processes ensure the formation of different metal compounds, their relationships, the strength of retention by soil particles, and, hence, the migration and accumulation of metals in soils.

The determination of compounds differently retained in the soil is an important problem in the fractionation of metal compound in both geochemical and ecological terms. Strongly retained compounds ensure the accumulation of heavy metals by soils. The migratory capacity of metals is related to their weakly retained compounds. They also control the water and biogenic migration of metals in soils. Therefore, these compounds are called potentially mobile or mobile compounds. Their presence in the soil in unoptimal amounts (excess or deficit) deteriorates the quality of natural waters and plants.

Fractionation procedures can be developed only on the basis of concepts of the formation mechanisms of metal compounds. Their involvement in a process depends on the element properties and soil-ecological conditions. For example, when the concentration of a component is lower than the solubility of its difficultly soluble compounds, sorption–desorption can be predominant processes responsible for its concentration; if the concentrations of components in the system correspond to the solubility product of their difficultly soluble compounds, the content of metal in the solution is mainly controlled by precipitation–dissolution of its compounds.

From the ecological point of view, of greatest importance is the separation of metal compounds in the soil into two groups of compounds: *strongly and weakly bound to soil components* (Tables 2.1, 2.2).

A group is a totality of metal compounds similar in binding strength to soil components and, hence, in migratory capacity and biological availability.

The group of strongly bound compounds includes the metals strongly fixed in the structures of primary and secondary silicate and nonsilicate minerals, as well as the metals in difficultly soluble compounds and stable organic and organomineral compounds.

The group of weakly bound compounds includes the metals retained on the surface of soil particles by organic and mineral soil components in the exchangeable and specifically sorbed states. It combines exchangeable, complex, and specifically sorbed metal compounds. Weakly bound compounds compose the group of heavy metals the most important in ecological terms, because they are first to get into plants and migrate to other adjacent environments. Hazardous ecological consequences of soil contamination with metals are related to these compounds.

The classification of heavy metal compounds in the soil is based on the similar strengths of their retention by solid phases and, hence, on the rates of release to soil solution. These properties determine the methods of extracting heavy metals from soil components and form basis for their fractionation scheme. These features are mutually related. The stronger the metal is bound to soil solid phases, the stronger the extractant required for its solubilization and vice versa.

In each group of metals, the binding strength of compounds also varies, and, hence, they can be fractionated according to this parameter (Tables 2.1, 2.2).

A fraction is a part of metal compounds from a group differing from other parts of the same group in the nature of binding to soil components. To assess the role of different metal fractions in the development of their mobility, their share in the group composition of metal compounds can be calculated.

Table 2.1. Compounds of metals strongly bound to soil components

Metal group	Metals strongly bound to mineral and organic soils components				
Metal-bound soil component	Mineral components				Organic components
Separated fraction	Residual			Fe and Mn sesquioxides	Organic matter
Fraction composition	Silicate minerals	Difficultly soluble metal salts		Crystallized nonsilicate minerals	Specific organic compounds
Metal form in the fraction	Metals in crystal lattices of primary and secondary minerals	Chemisorbed, occluded compounds	Difficultly soluble salt precipitates	Occluded compounds	Chelate compounds	Chemisorption complexes
Process	Isomorphic substitution, occlusion	Chemisorption, occlusion	Precipitation	Occlusion	Specific sorption	Complexation
Bond nature	Ionic, covalent, donor–acceptor, hydrogen	Ionic		No chemical bond	Coordination, donor–acceptor, hydrogen

Table 2.2. Compounds of metals weakly bound to mineral and organic soils components

Metal group	Metals weakly bound to mineral and organic soils components				
Metal-bound soil component	Mineral components				Organic components
Separated fraction	Specifically sorbed		Exchangeable		Complex
Fraction composition	On carbonates	On amorphous sesquioxides	Ion-exchange highly dispersed organic, mineral, and organomineral soil (SEC) components		Specific and nonspecific organic compounds
Metal form in the fraction	Surface (specifically sorbed) coordination compounds		Nonspecifically sorbed (exchangeable) surface compounds	Nonspecific sorption (ion exchange)	Surface complexes
Process	Specific sorption				Specific sorption
Bond nature	Donor–acceptor, hydrogen		Ion–ion, ion–dipole		Donor–acceptor, coordination, hydrogen

The main factors controlling the content of *weakly bound metal compounds* (water-soluble, exchangeable, complex, and specifically sorbed on the solid-phase surface) are adsorption–desorption equilibrium, ion exchange, and surface complexation. They are characterized by the high reaction rate and the low activation energy. The retention strength of heavy metals due to each reaction and the leading factors affecting their course are different. The leading factors are the surface size and quality (affinity to the element) for the sorption–desorption reactions, exchangeable sites (also characterizing the size and quality of the surface) for ion exchange reactions, pH for the precipitation–dissolution of metal compounds, and the presence of complexing agents for the formation of surface complexes. Extractants should be capable to enter into these reactions, i.e., ensure the transition of metals into solution from the corresponding compounds.

The main factors controlling the content of *strongly bound compounds* are chemisorption–desorption equilibrium, occlusion, isomorphic substitution, and (in contaminated calcareous soils) precipitation–dissolution of difficultly soluble salts. These processes need much energy; the release of metals strongly bound by soil particles proceeds with a low rate. The pH level and the concentration of ions participating in the reaction are the leading factors in the precipitation–dissolution of compounds on the surface of solid particles or as separate phases. The size and quality of the solid-phase surface and the structural organization of minerals can affect the process.

The distribution of metals between the pools of strongly bound and mobile compounds is controlled by a group of transformation processes (Fig. 2.1). The predominance of some heavy metal compounds depends on the composition and properties of soils, as well as natural and technogenic factors.

The mobility of metals in the soil can be assessed from the ratio between their groups and expressed in the form of mobility coefficient **Km**.

Km is the ratio of the weakly bound (WB) to the strongly bound (SB) metal forms in the soil:

Km = WB/SB

This method of assessing the mobility of heavy metals in the soil can serve as a criterion of soil contamination and possible translocation of metals into plants. It should be kept in mind that the availability of a metal for plants depends not only on the content of its weakly bound forms, but also on the physiological features of plants, their growing (soil-ecological) conditions, and the properties of the metal (its biophily).

Methodology of Studying Heavy Metal Compounds in Soils 29

| Strongly bound heavy metals in silicates, difficultly soluble salts, Fe and Mn (hydr)oxides, insoluble complexes with organic matter | ⇄ | **Processes:** dissolution–precipitation, complexation, oxidation–reduction, sorption–desorption, isomorphic substitution, and occlusion | ⇄ | Weakly bound heavy metals: organic and inorganic complexes, adsorbed on the surface of mineral and organic substances |

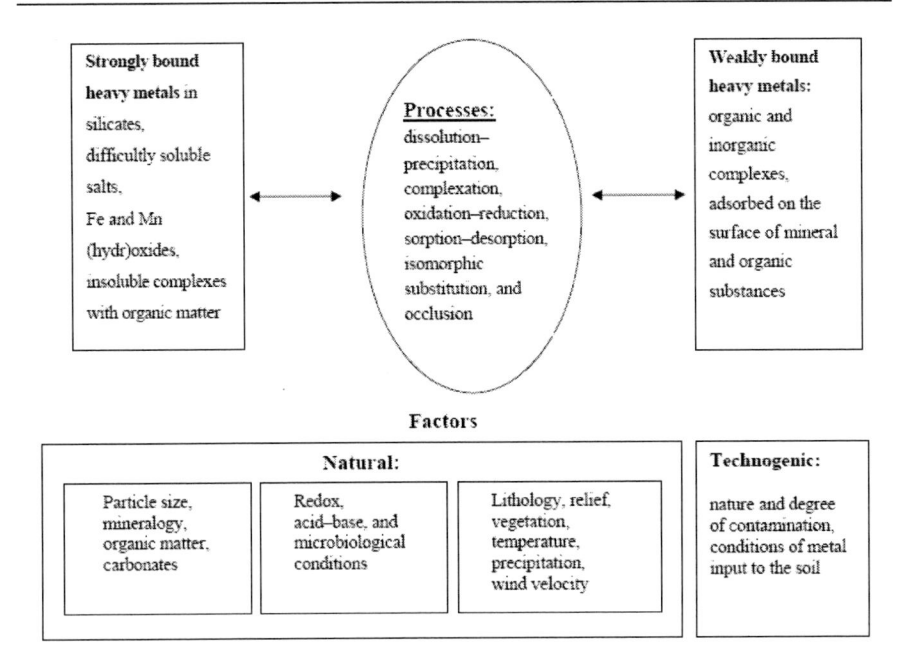

Factors

Natural:			Technogenic:
Particle size, mineralogy, organic matter, carbonates	Redox, acid–base, and microbiological conditions	Lithology, relief, vegetation, temperature, precipitation, wind velocity	nature and degree of contamination, conditions of metal input to the soil

Figure 2.1. Transformation of metal compounds in soil.

The approaches to determining the group composition of metal compounds in soils are similar to those widely used in pedological practice for other chemical elements. The methods of assessing the group composition of iron, aluminum, silicon, and phosphorus compounds and humic substances have been successfully used in soil studies over more than half a century. They are based on the solubility of their compounds. The ratio between the contents of isolated element compounds is indicative of soil-forming conditions, which confirms the efficiency of these approaches.

For example, L. O. Karpachevskii et al. (1972) and S. V. Zonn (1982) revealed tendencies in the formation of iron compounds in soils with different degrees of hydromorphism on the basis of differences in the solubility of iron oxides and hydroxides due to their degree of dispersion.

Tendencies in the formation of the group and fractional composition of humus in soils of different natural zones were revealed from the differences in the hydrolyzability of soil humic substances because of different binding strengths to the mineral matrix (Tyurin, 1937; Ponomareva, Plotnikova, 1968; Orlov, 1974). Information on the group and fractional composition of phosphorus (Chirikov, 1956; Ginzburg, Lebedeva, 1971), manganese (Schachschabel, 1956), and carbonates of different degrees of dispersion determined by the Drouineau–Galet method (Peterburgskii, 1968) are used for assessing soil fertility.

Groups of iron and aluminum compounds are extracted from separate soil samples (Theory and Practice of Soil Chemical Analysis, 2006). The content of group of these element compounds in the soil is determined using a combination of group and individual extractants and the summation or subtraction of the results. Analogous approaches can be used for the determination of the group composition of metal compounds.

The ratio of metal compounds strongly and weakly bound by soil components is dynamic in nature. The transformation of exogenic metals sorbed by the soil occurs in two stages: a relatively rapid and a slow transformation. At the first stage, the sorption of metals by the soil proceeds rapidly and is accompanied by the formation of their mobile compounds. However, no equilibrium is reached in a heterogeneous system even over a relatively long time period (Obukhov, 1989; Minkina et al., 2007). Only with time, the transformation of the metals sorbed occurs at the second stage, which results in a decrease in the content of their weakly bound compounds and an increase in the content of strongly bound compounds.

For example, the sorption of heavy metal ions by minerals involves three processes: surface adsorption, diffusion, and intrastructural fixation. Adsorption has the highest rate. Diffusion and especially intrastructural fixation of metals, which occurs through occlusion and isomorphic substitution, proceed slowly and result in the strongest fixation. Heavy metal compounds in these soil components cannot release into soil solution within a reasonably foreseeable time period and have no direct effect on the composition of mobile compounds. Our studies and available literature data (Obukhov, 1989) show that the content of water-soluble, exchangeable, and acid-soluble heavy metals decreases with time after the addition of metals to chernozemic soils.

In the soil, metals usually form no separate phase but are retained by macroelement compounds. The binding of metals to soil macrocomponents is differently manifested during their fractionation. According to an approach, fractionation is aimed at solubilizing carbon, iron, and aluminum compounds and determining the content of microelements in the resulting solutions. The nature of binding between microelements and the macrocarrier is ignored. Another approach implies the separation of heavy metal compounds on the basis of their bonds with macrocomponents, but without any indication to the macrocomponent composition. Different approaches disturb the logic of fractionation and frequently result in contradictory statements and superposition of results.

The interaction of heavy metal ions with soil components is due to different forces (Tables 2.1, 2.2): electrostatic (for oppositely charged ions), molecular (for neutral polar complexes and molecules), and chemical (mainly donor–acceptor). The donor–acceptor interaction of adsorbing ions with the surface functional groups (specific sorption) results in the formation of outer- and inner-sphere complexes surface (Pinsky, 1997). The presence of these complexes, along with the formation of polynuclear surface complexes, homogeneous precipitation, and diffusion in the lattice, was proved by EXAFS spectroscopy (Vodyanitskii, 2005b).

The identification and classification of metal compounds are rather conventional and related to the purposes of fractionation. Methods of determining these compounds are also of importance. Direct methods of determining heavy metal compounds in soils and revealing their binding to the soil matrix have been actively developed recently, which provide valuable information on the sample composition. The X-ray synchrotron technique is most efficient.

However, X-ray diffractometry is frequently low efficient, because it cannot reveal small amounts of these metals and relate them to the carrier phases. Transmission electron microscopy, which is accompanied by the microdiffraction of electrons, can reveal the substitution of heavy metal ions for iron and manganese ions in their oxides (Vodyanitskii, 2005b). However, it implies the disintegration of soil sample, which disturbs the actual soil phase ratios in some cases.

Synchrotron accelerators, which impart energy to charged particles, have found recent use in soil science. Such works are being conducted only in a few world centers: the European Synchrotron Radiation Facility (ESRF) in France, as well as the Advanced Light Source (ALS) and the Stanford Synchrotron Radiation Lightsource in the United States. These studies were performed for a small group of soils (predominantly acid soils).

Liquid extraction techniques are still widely used for the solubilization of different metal compounds, although the first of them were developed as early as the 1960s. The most common are different versions of sequential (presumably selective) extraction. The reagents used should meet two main requirements: they should ensure the completeness and maximum selectivity of extraction of specific metal compounds. However, the absolute compliance with these requirements is hardly attainable for such a complex heterophase system as the soil. Heavy metal ions are fixed by different soil components through different reactions; therefore, the extraction fractionation of their compounds is not selective. This imposes principal limitations on all extraction techniques. Authors most frequently follow Tyurin, who called the fractions of humic substances separated according to his scheme "presumably bound" to specific components.

Many reagents are used for extracting different heavy metal compounds. Since the 1960s, numerous schemes have been proposed for the sequential fractionation of microelement compounds from the soil (Le Rich, Weir, 1963; Chao, 1972; McLaren, Crawford, 1973; Zyrin, Obukhiv, Motuzova, 1974; Andersson, 1976; Gatenhouse, Russel, van Moort, 1977; Quy, Chakrabarti, Bain, 1978; Tessier et al., 1979; Shuman, 1983; Gibson, Farmer, 1986; Miller et al., 1986; Zien, Brummer, 1991; Whalley, Grant, 1994; Ladonin, 1995; Einax, Nischwitz, 2001; Gleyzes, Tellier, Astruc, 2001; Ahumada et al., 2003; Mossop, Davidson, 2003; Motuzova, Aptikaev, Karpova, 2006; Zemberyova, Bartekova, Hagarov, 2006; Pueyo et al., 2008). The Tessier procedure has found wide use for the fractionation of soil metal compounds (Tessier et al., 1979). Reviews of reagents used for determining different metal forms were reported by Kuznetsov and Shimko (1990), Sadovnikova (1997), Ponizovskii and Mironeneko (2201), and Ladonin (2002) (Table 2.3).

All the above fractionation techniques (Table 2.3) are based on the assumption about the relatively selective effect of the proposed reagents on the macrocomponents to which the metals are presumably bound.

The selectivity of extractants can be controlled by comparing the composition of extracted metal compounds with the parameters of their composition determined by independent instrumental methods (Scheinost, Kretzschmar, 2001; Dahn R at al., 2002; Lamoureux at al., 2001; Scheinost, Kretzchmar, Pfister, 2002; Vodyanitskii, 2005a, 2008). For example, the synchrotron radiation method provided information on the forms of lead and zinc compounds in acid soils, which was compared to the results of their fractionation by different methods (Table 2.4).

The selection of extractant for each fraction of heavy metal compounds is based on the concepts of retention mechanisms of these compounds by soil components (Fig. 2.2).

Table 2.3. Extractants of different heavy metal compounds from the soil (Ponizovskii, Mironenko, 2001; Ladonin, 2002; Vodyanitskii, 2005a)

Authors	Mobile (exchangeable)	Mobile (specifically sorbed on carbonates)	Bound to organic matter	Bound to Fe and Mn oxides and hydroxides	Bound to aluminosilicates
Le Richie, Wear, 1963	CH_3COONH_4, pH 7		Tamm's reagent under UV irradiation		Residue after the preceding treatments
Crimme, 1967	Not determined		0.5 N NaOH + 0.1 M EDTA; 0.2 M ammonium oxalate, pH 3	0.4 M ammonium oxalate, pH 3.6 in the presence of Zn	Residue after the preceding treatments
Taranovskii, Sochilina, 1969	1 N NH_4Cl in 70% CH_3COOH (pH 6.5–6.8)		Not determined	1 N CH_3COOH, 6 h at 100°C	Residue after the preceding treatments
Titova, 1970	CH_3COONH_4, pH 4		1 N HCl after treatment with 30% H_2O_2*0.05M EDTA	Mehra–Jackson reagent	In clay minerals, treatment with 20% HCl; in primary minerals, residue after the preceding treatments
Motuzova, 1973	1 N CH_3COONH_4, pH 4.8		1 N H_2SO_4 after treatment with 30% H_2O_2	Tamm's reagent under UV irradiation	In the fraction <0.01 mm of clay minerals; in the fraction >0.01 mm of primary minerals
McLaren, Crawford, 1973	0.05 M $CaCl_2$	2,5% CH_3COOH	1 M $K_4P_2O_7$, pH 11	Tamm's reagent under UV irradiation	Residue after the preceding treatments
Tessier, Campbell, Bisson, 1979	1 M $MgCl_2$, pH 7	1 M CH_3COONa + CH_3COOH, pH 5	H_2O_2 + 0.02 M HNO_3, pH 2, then 3.2 M CH_3COONH_4 in 20% HNO_3	0.04 M NH_2OH–HCl in 5% CH_3COONH_4	HF + $HClO_4$, heating and evaporation to dry

Table 2.3. (Continued).

Authors	Mobile exchangeable	specifically sorbed on carbonates	Bound to organic matter	Bound to Fe and Mn oxides and hydroxides	Bound to aluminosilicates
Zeien, Brummer, 1989	1 M NH$_4$NO$_3$	1 M CH$_3$COONH$_4$, pH 6	0,025 M NH$_4$-EDTA, pH 4,6	Mn: 0.1 M NH$_2$OH–HCl + 1 M CH$_3$COONH$_4$, pH 6; Amorphous Fe: 0.2 M (NH$_4$)$_2$C$_2$O$_4$, pH 3.25, in dark; Crystallized Fe: 0.1 M ascorbic acid + 0.2 M (NH$_4$)$_2$C$_2$O$_4$, refluxing	Refluxing of residue in a mixture HNO$_3$ + HClO$_4$, 3:1, dissolution of precipitate in 5 M HNO$_3$
BCR, 1994	Not determined	0.11 M CH$_3$COOH, pH 3	27% H$_2$O$_2$; 1 M CH$_3$COONH$_4$, pH 2	0.1 M NH$_2$OH–HCl, pH 2	HNO$_3$ + HCl
Ladonin, 1995	0.05 M Ca(NO$_3$)$_2$	2.5% CH$_3$COOH	0.1 M EDTA, 1 M K$_4$P$_2$O$_7$, pH 11	Tamm's reagent under UV irradiation	Residue after the preceding treatments
Salim, Miller, Howard, 1996	1 M MgCl$_2$	1 M CH$_3$COONa	0.1 M K$_4$P$_2$O$_7$	0.1 M NH$_2$OH–HCl + 4.37 M CH$_3$COOH	Residue after the preceding treatments

Table 2.4. Pb and Zn compounds detected in soils by the X-ray synchrotron technique (Manceau et al., 2000, 2002; Scheinost et al., 2002; Pierzynski et al., 2008) and extracted by the Tessier and BCR methods

Extracted compounds	Minerals
Humates of heavy metal	Pb-humate
Fe and Mn oxides	
Pb-contained	Pb on goethite
	Pb on bernessit
	Plumbojarosite Pb[Fe3(SO4)2(OH)6]2
	Plumboferrite [Pb2Mn0,2Mg0,1Fe10,6O18,4]
	Magnetoplumbite [PbFe$_6$Mn$_6$O$_{19}$]$_2$
Zn- contained	Franklinite Zn$_2$FeO$_4$
	Zn- magnetite
	Zn- goethite
	Zn- feroxyhyte
	Zn- lithiophorite
Carbonates	
Pb- contained	Cerrusit Pb$_3$(CO$_3$)$_2$
	Hydrocerrusit Pb$_3$(CO$_3$)$_2$(OH)$_2$
Zn- contained	Zn-hydrotalkit (?) Zn$_3$Al(OH)$_8$(CO$_3$)$_6$
	Hydrozincite (?) Zn$_5$(OH)$_6$(CO$_3$)$_2$
Phosphates are not found	Pyromorphite Pb$_5$Cl(PO$_4$)$_3$
Silicates are not found	Willemite Zn$_2$SiO$_4$
	Kerolite Si$_4$(Mg$_{2.25}$Zn$_{0.75}$)O$_{10}$
	Zn-hectarit
	Zn- montmorillonite
	Gemomorfit Zn$_4$(Si$_2$O$_7$)(OH)$_2$

The determination of strongly bound metal compounds presumes the decomposition of the soil component to which the metal is bound. The methods of solubilizing metal compounds strongly retained by silicate minerals should ensure the complete digestion of the sample. This is reliably reached by two methods: (a) the fusion of sample with different fluxes (carbonates, borates, etc.) at about 1000°C and (b) the decomposition of sample under heating with a mixture of mineral acids (H$_2$SO$_4$, HNO$_3$, HClO$_4$) and HF, whose decomposing effect is ensured by the formation of volatile SiF$_4$. For the soils developed on deeply weathered rocks and the strongly contaminated soils with the predominance of exogenic metal compounds (e.g., oxides), the decomposition of sample by a mixture of mineral acids (H$_2$SO$_4$, HNO$_3$, HClO$_4$) at heating under pressure can be efficient.

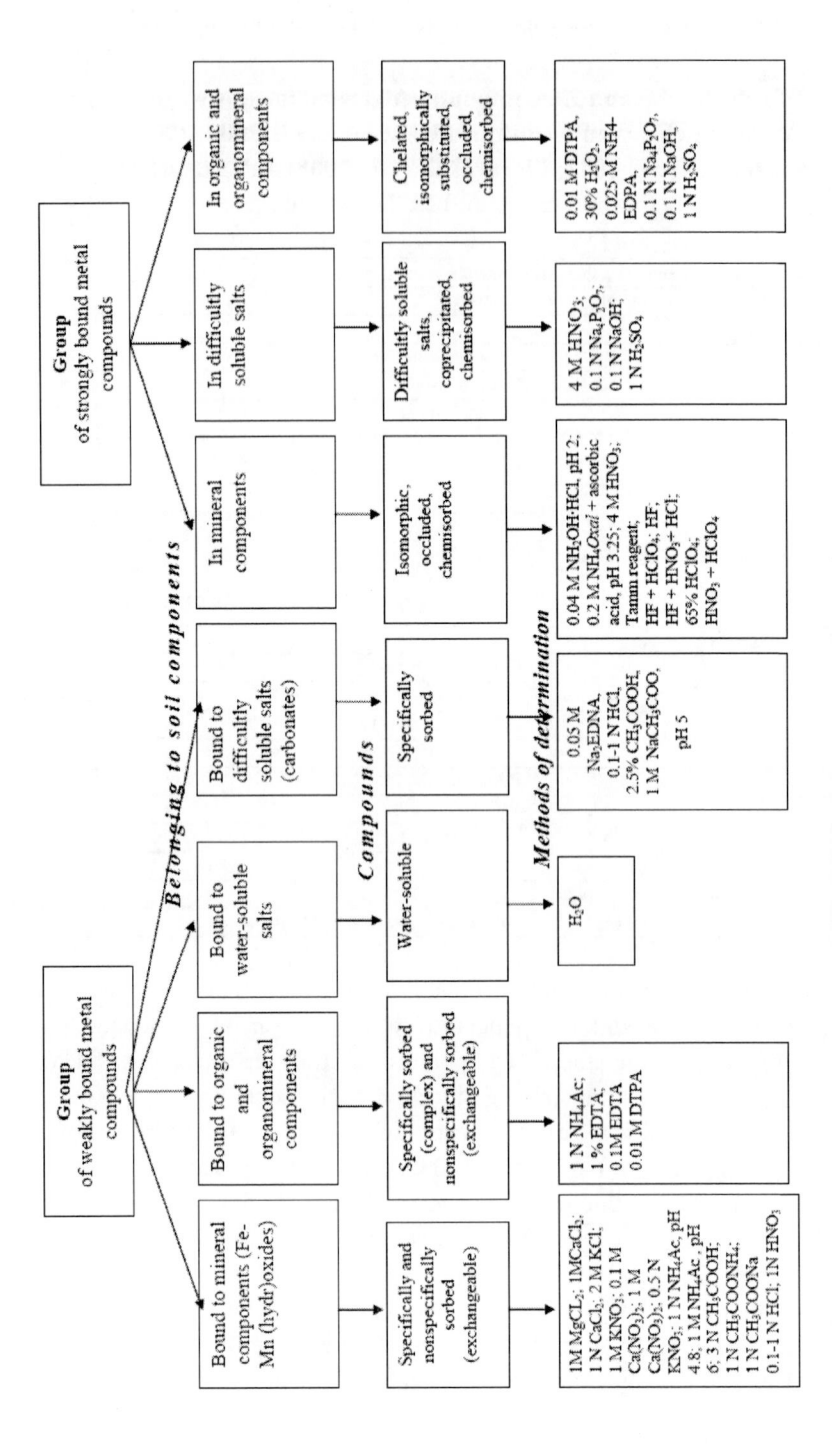

Figure 2.2. Heavy metal compounds in soils and methods of their determination.

Two groups of essential pedogenic components (Fe, Al, and Mn nonsilicate compounds and organic substances) strongly retain heavy metals. In saline soils, these can also be carbonates and other salts. Reductants, diluted solutions of acids (H_2SO_4, HNO_3, HCl), are used for the complete solubilization of Fe and Mn oxides and hydroxides and metals retained by them. Oxidants, alkaline solutions, are used for the complete solubilization of organic substances and retained metals.

The extraction of nonspecifically sorbed metals is based on their displacement from the soil exchange complex by an excess of an exchangeably sorbed cation. In the reagents used for the displacement of exchangeable heavy metal ions in the soil, the displacing cation should meet the following requirements (Gorbatov, Zyrin, 1988): (1) it should have a high energy of introduction into the SEC; (2) it should neither specifically interact nor form precipitates or complexes with soil components; (3) its ionic radius should be similar to that of the displaced cation; and (4) it should not interfere with the following analytical determination of the extracted ion.

The strict observation of all these requirements is not always feasible, especially at the simultaneous determination of several elements. In practice, different extractants are frequently used, which hampers the comparison of the results (Fig. 2.2).

The methods of determining specifically sorbed metals are based on two different approaches. An approach uses the displacement of the sorbed metal ions by protons (McLaren, Crawford, 1973). The other approach is based on the displacement of specifically sorbed metal by ions of another metal, which is present in excess and can be specifically sorbed by the soil (Ramamoorthy, Rust, 1978). The specific binding of metal ions to soil components can be due to different mechanisms; therefore, specifically sorbed heavy metals are solubilized using reagents with different extraction capacities. Diluted acids and buffer solutions are most frequently used in soil studies (Fig. 2.2).

The content of metals specifically sorbed by soil organic matter (heavy metal complexes) is determined using different complexing agents like DTPA, EDTA, and EDNA. The main condition of their selection is the capacity to form complexes with metal ions that are more stable than those binding them to soil organic matter.

The most important problems arise at the attempt to more selectively separate metal compounds retained by nonsilicate ions and soil organic matter. In the nature, they are mainly interrelated; their artificial separation under laboratory conditions pursues purely research goals. Organic and mineral compounds solubilized during the fractionation of strongly bound heavy metal compounds (Fig. 2.2) bear sorption centers of different nature and, hence, contribute to the retention of mobile metal compounds. However, they are not differentiated during the fractionation of weakly bound metal compounds.

Thus, the heavy metal compounds determined in the soil by extraction methods are rather conventional. However, they give an idea of differences in the strength of their retention by soil components and, hence, mobility. The increase in the information value of data obtained by the extraction methods has ecological significance. This can be attained by using, along with the sequential fractionation of metal compounds from the soil, their parallel extraction and a combination of these two methods.

2.2. DETERMINATION OF HEAVY METAL COMPOUNDS BY PARALLEL EXTRACTION

In the practice of soil studies, individual extractants of metal compounds or combinations of individual extractants used separately (parallel extractants) have been used over a relatively long time period. Information on the content of metals weakly bound by soil components, i.e., potentially capable of migration, should be first acquired using individual extractants. The individual reagents recommended by Peive and Rin'kis in 1958 for extracting mobile compounds of microelements (Cu, Zn, Mn, Co, B, Mo) most valuable for living organisms from soils present a demonstrative example. These methods found wide use. Unique data on the supply of soils in the European regions of the USSR with mobile microelement compounds were obtained using these methods (Trace Elements in the Soils of the Soviet Union, 1973). However, no indices were available on the specific form of occurrence of mobile (or weakly bound) microelements in the soil.

Ammonium acetate buffer solution (1 N AAB) with different pH levels was a common group extractant for mobile microelement compounds; the solution with pH 4.8 was most frequently used according to the Krupskii–Aleksandrova method (1964). Only in the last decade of the 20th century, an idea that the content of AAB-extractable metals characterizes the total reserve of their exchangeable compounds in the soil was established.

A mixed extractant (1 N AAB + 1% EDTA) proposed by Crimme (1967) was recognized by investigators of soil microelements, because it allowed accounting for the contribution of not only exchangeable, but also complex compounds of metals, to their total reserve.

Studies of metal-contaminated soils developed in the 1980s contributed to the extension of methods for acquiring prompt information on the environmental hazard of polluted soils. Such information can be derived from the content of metals in acid extracts (1 N HNO$_3$, 1 N HCl). These extractants solubilize both the metals weakly bound to the soil matrix and the technogenic metal oxides arrived in the soil but still not involved in transformation.

The ratio of metal compounds extracted by the above reagents is shown in Fig. 2.3. The presumable partial superposition of metal compounds extracted by three individual (parallel) reagents offers opportunity to calculate the content of some metals by subtracting the amount of less strongly retained compounds (extracted by a weaker reagent) from the amount of more strongly retained compounds (extracted by a more aggressive reagent).

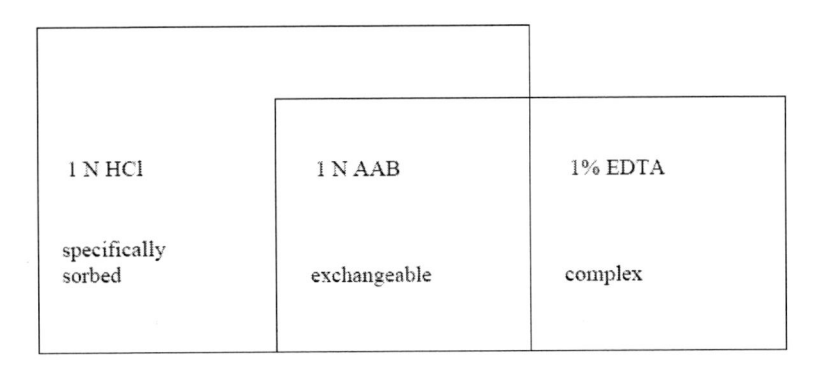

1 N HCl	1 N AAB	1% EDTA
specifically sorbed	exchangeable	complex

Figure 2.3. Extraction of weakly bound metal compounds by individual reagents (parallel extraction).

These reagents are recommended for characterizing the integrated state of mobile heavy-metal compounds in the soil by the Solov'ev method (Laboratory Manual on Agricultural Chemistry, 1989):

(1) Metal compounds extractable by 1 N ammonium acetate buffer solution (CH$_3$COONH$_4$ or AAB) with pH 4.8 (soil : solution ratio of 1 : 5, extraction time 18 h) characterize the actual reserve of mobile metal forms in the soil. The 1 N CH$_3$COONH$_4$ solution has a mixed effect on the soil. Under its effect, heavy metal ions can pass from the soil into solution due to the exchange for NH$_4^+$ and H$^+$ ions and hydrolysis of some readily hydrolyzed compounds under the effect of hydrogen ions. The contribution of the latter ions is small: the content of soluble compounds is no more than 1–2% of the total amount for most heavy metals (Suslina et al., 2006).

The single treatment of soil with a 1 N CH_3COONH_4 solution does not completely displace all exchangeable Pb and Zn ions (Minkina et al., 1998), because no more than 50% of exchangeable alkali cations, which are retained in the SEC much less strongly than heavy metal ions, are extracted in this case (Ponizovskii, Polubesova, 1990). On the other hand, 1 N CH_3COONH_4 can solubilize metal ions not only during ion exchange, but also due to their inclusion in acetate complexes.

Thus, the least strongly retained metal compounds are predominantly extracted by 1 N CH_3COONH_4 (Soon, Bates, 1982; Ladonin, 2002). These metal compounds are bound to different soil components by nonspecific sorption (Fig. 2.2) and are most frequently called exchangeable (Laboratory Manual on Agricultural Chemistry, 1989; Sadovnikova, 1997; Dabakhov, Solov'ev, Egorov, 1998; Suslina et al., 2006).

(2) The mixed reagent (1 N CH_3COONH_4 + 1% EDTA) with pH 4.8 (soil : solution ratio of 1 : 5, extraction time 18 h) containing a more active complexing agent than the acetate ion can solubilized, along with exchangeable forms, metals bound to complexes with organic substances. It is suggested that extractable metals are relatively weakly bound to organic matter (McLaren, Crawford, 1973). The mixed reagent (1 N CH_3COONH_4 + 1% EDTA) has a higher extracting capacity with respect to metal ions than 1 N CH_3COONH_4 (Nikityuk, 1998). A close correlation (r = 0.94–0.95) is observed between the contents of metals in the extract and in plants, which points to the availability of thus extractable metal compounds for plants.

The content of metal complexes in the soil can be calculated from the difference between the metal contents extracted by the mixed reagent (1 N CH_3COONH_4 + 1% EDTA) and 1 N CH_3COONH_4 (Sadovnikova, 1997).

(3) The analysis of extracts by diluted (1 N) acids (HCL, HNO_3, H_2SO_4) is conventionally used for characterizing the ecological status of soils. The metal compounds extractable by these reagents are called acid-soluble. Acids solubilize exchangeable metal ions and metals specifically sorbed by carbonates and Fe–Mn (hydr)oxides. The latter include finely dispersed (amorphous) iron compounds more soluble than coarsely dispersed ones. There is a wide experience of using diluted acids for extracting mobile iron compounds: I. K. Kazarinova-Oknina proposed 0.1 N H_2SO_4 and K. V. Verigina proposed 1 N H_2SO_4 (Arinushkina, 1970) for the extraction of mobile iron compounds from soils. The content of acid-soluble metal compounds presumably characterizes the potential reserve of their mobile compounds. A 1 N HCl extract is prepared at a soil : solution ratio of 1 : 10 and a time extraction of 1 h.

The effects of 1 N CH$_3$COONH$_4$ with pH 4.8 and 1 N HCl on mobile heavy-metal compounds are different (Minkina et al., 2008a). The 1 N CH$_3$COONH$_4$ ensures (a) the displacement of exchangeable metal cations by ammonium ions and protons (the displacing function is most due to ammonium, because of the high concentration of the buffer mixture, 1 mol/l) and (b) the binding of the displaced metal ions into soluble acetate complexes.

When 1 N HCl is used as an extractant, its protons are consumed for the displacement of exchangeable metal cations and those specifically sorbed by Fe and Mn hydroxides, as well as for the neutralization of carbonates and the release of metal ions retained by them. Therefore, is may be suggested that the difference between the contents of metal in the 1 N HCl and 1 N CH$_3$COONH$_4$ extracts corresponds to the amount of metals bound to carbonates and Fe–Mn hydroxides.

The calculation of complex and specifically sorbed heavy-metal compounds is based on the assumption of their additivity. To verify this assumption, two reagents (1 N AAB + 1% EDTA and 1 N AAB) were used for the parallel and sequential extraction of metal compounds. An analogous experiment was performed with other two reagents: 1 N AAB and 1 N HCl. A good agreement was found between the results of sequential extraction of metals and those obtained by calculation from the difference between their contents in parallel extracts (Table 2.5). The differences were statistically insignificant and lied within the experimental error.

The content of metals in strongly bound compounds with organic and mineral components was calculated from the difference between the total metals in the soil and their weakly bound forms.

The calculation method of determining different compounds was also used by other authors. M. N. Khoroshkin and B. M. Khoroshkin (1979) calculated the content of strongly bound, difficultly available metal compounds from the difference between the total metal in the soil and its available (salt and acid) forms (Table 2.6). V. G. Unguryan and A. M. Kholmetskii (1978) calculated the content of strongly bound and almost insoluble heavy metals from the difference between their total content in the soil and their mobile forms, as well as the content of exchangeable and readily soluble metals from the difference between their mobile and water-soluble forms. In the study of A. I. Obukhov and M. A. Tsaplina (1990), strongly fixed forms of a metal were assessed from the difference between its total content and concentration in an ammonium acetate extract.

Table 2.5. Comparison of metal contents determined by parallel and sequential extractions, mg/kg (n = 9)

Dose of metal, mg/kg	Complex forms			Specifically sorbed forms		
	by the difference [(EDTA + AAB) − AAB]	Sequentially (EDTA after AAB)	LSD$_{0.95}$	by the difference (HCl − AAB)	Sequentially (HCl after AAB)	LSD$_{0.95}$
Cu						
0	0,2	0,3	Insignificant differences	1,9	1,5	0,2
55	3,1	3,5	0,2	18,6	17,1	0,5
100	10,8	10,0	0,5	25,0	23,1	0,9
Pb						
0	0,2	0,4	0,1	2,3	2,0	0,1
55	4,4	4,0	0,2	5,6	4,9	0,4
100	9,4	12,5	0,9	19,1	26,3	1,5
Zn						
0	0,3	0,5	0,1	6,6	5,9	0,3
55	1,9	2,5	0,3	46,2	43,5	2,0
100	4,0	5,6	0,4	67,2	62,4	1,7
300	20,7	18,6	0,4	90,7	99,1	4,4

Table 2.6. Copper and zinc in soils of Rostov oblast, % of their total content (M.N. Khoroshkin, B.M. Khoroshkin, 1979) (parallel extractions)

Extracted compounds	Cu	Zn
Water extraction, the most accessible	trace	trace
Salt extraction, easily accessible	0,8	0,9
Acid extraction, accessible	14	15
Strongly bound, hardness accessible	86	84

In another approach, the sequential treatment with a series of group extractants is used. These can be AAB and 1 N HCl (Suslina et al., 2006; Strnad et al., 2006) (Tables 2.7 and 2.8, respectively); EDTA and 1 N HCl (Fedorov, 2007); Ca(NO$_3$)$_2$, AAB, and 1 N HCl (Popovicheva, 1988); and 1 M NH$_4$Cl, AAB, and 0.1 M NaOH (Karpova, 2006).

Table 2.7. The content of exchangeable (AAB extraction) and specifically sorbed (HCl extraction) compounds of copper, zinc and lead in soils, % of their total content (Suslina et al., 2006) (sequentially extraction)

Metal	Sandy soddy-podzolic soil		Platic soil	
	AAB	1N. HCl	AAB	1N. HCl
Cu	32,8	63,0	5,7	43,5
Pb	67,7	29,6	20,2	38,3
Zn	57,1	11,9	22,7	34,3

Table 2.8. The content of exchangeable and specifically sorbed (HCl extraction) compounds of copper and lead, % of their total content (Strnad et al., 1991) (sequentially extraction)

Metal	Extracted compounds	
	exchangeable	specifically sorbed
No added metal		
Cu	0,30±0,08	0,69±0,38
Pb	0,15±0,05	6,36±0,22
Dose of metal, 100 mg/kg		
Cu	28,38±4,80	53,59±16,80
Pb	20,64±2,75	73,24±5,40

In the proposed calculation method, available data on the fractional composition of metal compounds in different soils can be used and expanded with the calculated parameters of metal contents in other (nonextracted) fractions.

The interpretation of any fractionation method's results is rather conventional. The above groups of mobile metal compounds in soils form a continual series in accordance with increasing stability. They are related by diverse intermediate forms differing in formation rate. The method of calculating the content of metals in separate groups from the difference between the separate extracts also includes some conventionality. This is related to the fact that the displacement of the same heavy-metal forms by different agents follows different mechanisms, which can result in disagreement between the contents of extracted compounds. Soil properties also affect these processes.

2.3. DETERMINATION OF HEAVY METAL COMPOUNDS USING A COMBINED FRACTIONATION SCHEME

The analysis of various compounds of chemical elements in soils is often performed with the use of sequential fractionation. Diverse schemes of fractionation are based on the same approach: heavy metal compounds are separated into two groups: compounds weakly bound and those strongly bound to the soil matrix.

Table 2.9. Sequential and parallel fractionation of metals

Metal compounds	Extractant	Soil : solution ratio	Extraction conditions
Sequential fractionation (Tessier et al., 1979)			
Exchangeable	1 M $MgCl_2$, pH 7.0	1:8	shaking at room temperature for 1 h
Bound to carbonates	1 M $NaCH_3COO$, pH 5.0 (with CH_3COOH)	1:8	shaking at room temperature for 5 h
Bound to Fe, Al, and Mn (hydr)oxides	0.04 M $NH_2OH \cdot HCl$ in 25% CH_3COOH	1:20	heating at 96±3°C under periodical shaking for 8 h
Bound to organic matter	0.02 M HNO_3 + 30% H_2O_2, pH 2.0 (with HNO_3), then 3.2 M NH_4CH_3COO in 20% HNO_3	1:20	heating at 85±2°C under periodical shaking for 5 h
Residual fraction	$HF+HClO_4$, then conc. HNO_3	1:25	evaporation
Parallel extraction (Solov'ev, 1989)			
Exchangeable	NH_4Ac, pH 4.8	1:5	shaking for 1 min and settling for 18 h
Exchangeable + complex	1% EDTA in NH_4Ac, pH 4.8	1:5	shaking for 1 min and settling for 18 h
Exchangeable + specifically sorbed	1 N HCl	1:10	shaking for 1 h

As was noted in Section 2.1, weakly bound heavy metal compounds are separated in accordance with the mechanism of their interaction with the soil components; thus, exchangeable, complex, and specifically adsorbed compounds can be distinguished. In order to determine the content of heavy metals strongly bound by the same soil components, they are solubilized together with their carriers (organic substances and nonsilicate compounds of Fe, Al, and Mn). The content of elements retained by silicate minerals characterizes the composition of the residual fraction. The metal compounds determined by this method are referred to as "presumably bound" to the above soil components, because the concepts of the solubility of natural microelement compounds in particular extracts are still incomplete (Motuzova, Degtyareva, 1991).

The methods of sequential and parallel fractionation of metal compounds are widely used separately, but they can form basis for their combined use (Minkina et al., 2008b). It is advisable to use the Tessier scheme of sequential fractionation (Tessier et al., 1979) in combination with the parallel extraction using individual reagents by the Solov'ev method (Table 2.9).

The Tessier procedure of fractionating metals is largely similar to other methods used for the analysis of soils, sediments, and bottom sediments. This procedure ensures the separation of five fractions of heavy metal compounds:

1. The exchangeable fraction contains metal ions mainly retained by electrostatic forces on the surface of clay and other minerals, organic substances, and amorphous compounds with the low pH values of zero charge. The fraction is affected by the ionic composition of soil solution. The exchangeable forms of heavy metals extractable from soils by neutral salt solutions mainly correspond to adsorbed forms weakly bound to the soil matrix (Pinsky et al., 1989).
2. The carbonate-bound fraction contains heavy metal ions specifically sorbed on Ca and Mg carbonates. Partial dissolution of metal phosphates is also possible (Kuhn, 1996).
3. The hydroxide-bound fraction contains heavy metal ions occluded by amorphous Fe and Mn hydroxides or adsorbed on their surface. Metals from organic complexes and amorphous sulfides can be released during extraction.
4. The organic matter-bound fraction contains heavy metal ions retained by organic substances or organomineral compounds. Sulfides can undergo partial decomposition.
5. The residual fraction contains heavy metal ions strongly fixed in the crystal lattices of primary and secondary minerals. This fraction can also include metals from stable sulfides and, in small amounts, stable organomineral substances.

The extraction capacity of individual reagents was discussed above (Section 2.1).

After each extraction step, the liquid and solid phases are separated by centrifugation. To decrease the effect of the multicomponent composition of extracts on the determination of their metal content by AAS, the standard addition method was used. The corresponding extracts from background soil samples were used for the reference scale.

All schemes of metal fractionation are not free from shortcomings: the separation of soil into components is accompanied by a shift of soil equilibrium and redistribution of metals in the soil (Vodyanitskii, 2005a); the exhausting extraction of analyte metal compounds is not involved (Perelomov, Pinsky, 2003); the affinity of metals to Mn oxides is not revealed (Vodyanitskii, 2005b); the reagents used are not selective.

The principal problem in the use of different schemes for the fractionation of soil metal compounds is that they insufficiently reflect the effect of diverse soil components retaining metal compounds. However, the accumulated experimental data indicate a significant differentiation of typomorphic chemical elements in soils. They show that the diversity of soil organic substances (Tyurin, 1937; Orlov, 1974); oxides of iron (Zonn, 1982), aluminum (Zonn, Travleev, 1992), and manganese (Vodyanitskii, 2005b); and exchangeable ions of SEC (Pinsky, 1992) reflects the effect of soil-forming conditions. These conditions obviously affect the state of metals bound to the above components and, hence, the binding strength of metals in the soils of natural and technogenic landscapes.

The presumable participation of organic and mineral soil components in the formation of strongly and weakly bound metal compounds is shown in Fig. 2.2. The effect of essential soil compounds (exchangeable and bound to organic matter; nonsilicate Fe, Al, and Mn minerals; and carbonates) during the fractionation of metal compounds should be taken into consideration. This problem cannot be solved by the sequential or parallel fractionation of metal compounds used separately. However, the concepts of extraction mechanisms of the reagents used in these schemes allow the determination of a number of metal compounds, which were not indicated in the fractionation methods, by subtracting the contents of presumed components from the total contents of some fractions. Similar operations are performed during the fractionation of iron and manganese compounds and organic substances.

On this basis, a combined scheme was proposed for the fractionation of soil metal compounds (Table 2.10):

Table 2.10. Combined fractionation scheme of heavy metals

Parameter	Determination method — experimental	Determination method — calculation (from the difference of contents in extracts)
1. Exchangeable metal forms:		
Total	1 N AAB, pH 4.8	
Readily exchangeable	1 M MgCl$_2$	
Difficultly exchangeable		Difference 1 N AAB – 1 M MgCl$_2$
2. Metals bound to carbonates and occurring as separate phases:		
Total	No method	
Weakly bound (specifically sorbed)	1M NaCH$_3$COO, pH 5	
Strongly bound (coprecipitated, occluded, chemisorbed, precipitates of low-soluble heavy metal compounds)	No method	
3. Metals bound to Fe, Al, and Mn nonsilicate compounds:		
Total	0.04 M NH$_2$OH·HCl	
Weakly bound (specifically sorbed)		Difference (1 N HCl – 1 N AAB) – 1 M NaCH$_3$COO
Strongly bound (occluded)		Difference 0.04 M NH$_2$OH·HCl – (1 N HCl – 1N AAB – 1 M NaCH$_3$COO)
4. Metals bound to organic matter:		
Total	30% H$_2$O$_2$	
Weakly bound (complexes)		Difference (1% EDTA in 1 N AAB) – 1 N AAB
Strongly bound (chelates)		Difference 30% H$_2$O$_2$ – 1% EDTA
5. Metals strongly bound to silicates	HF+HClO$_4$ extract from the residual soil fraction (after all extractions)	Difference between the total element content in the soil and the total content of all fractions (except the residual fraction)

(1) *Fractionation of exchangeable metal compounds.* Extraction with 1 N MgCl$_2$ recommended by Tessier characterizes the total content of water-soluble metal compounds and readily exchangeable metal ions in soils.

However, the reserve of exchangeable metal ions in the soil is significantly larger than the values found by this method because of the diversity of exchange centers. In the national and global practice (Wilde, Voigt, 1955; Arinushkina, 1961), their total content is determined in an ammonium acetate buffer solution (1 N CH$_3$COONH$_4$ or AAB), as was proposed by Pryanishnikov as early as in 1913 (Gedroits, 1955).

The extraction capacity of AAB with respect to metals is higher than that of an MgCl$_2$ solution. The presence of acetate ions increases the capacity of AAB to extract exchangeable ions from the soil. The content of metals in the AAB extract is almost always higher in comparison with that in the MgCl$_2$ extract (Zhideeva et al., 2002; Perelomov, Pinsky, 2003; Gordeeva, Belogolova, 2007). This is especially the case for the contaminated soils containing increased amounts of heavy metal ions in the SEC. The difference between the metal contents in the 1 N CH$_3$COONH$_4$ and 1 M MgCl$_2$ extracts can be indicative of the content of difficultly exchangeable metal ions in soils.

(2) *Fractionation of metal compounds bound to carbonates and forming own compounds.*

Heavy metals bound to carbonates and other difficultly soluble compounds and the methods of their determination in soils are of particular interest. There is more than one opinion concerning the forms of metals in difficultly soluble soil components.

At low concentrations, divalent metal cations are first adsorbed on the calcite surface. Then, they can be included into the calcite lattice by coprecipitation during recrystallization (Franklin, Morse, 1983; Kornicker et al., 1985; Pingitore and Eastman, 1986; Reeder and Prosky, 1986; Reeder and Grans, 1987; Davis et al., 1987; Zachara et al., 1988). It was shown that the surface of CaCO$_3$ has a high affinity for Cd ions (Papadopoulos, Roweli, 1988). The addition of 60 mg of CdO into suspensions from chernozem and sierozem resulted in the formation of solid cadmium carbonate (Gorbatov, 1988).

A dispersed precipitate of metal carbonate or basic carbonate can be formed on the surface of calcite in the presence of an appreciable amount of water (McBride, 1979, 1980; Glasner and Weiss, 1980; Kaushansky and Yariv, 1986; Papadopoulos and Roweli, 1988, Zachara et al., 1988). Papadopoulos and Roweli (1988) found that at a high Cd concentration, CdCO$_3$ precipitates were predominant on calcite and the precipitation was slow. Stipp et al. (1992) showed that Cd was first adsorbed on the $CaCO_3$ surface and then slowly diffused into the inner spheres of a particle to form a mixed $MeCO_3 + CaCO_3$ precipitate. According to Zachara et al. (1988), at a high metal concentration (10^{-5} to 10^{-4} M/l), equilibrium solutions were saturated with the dispersed solid sorbent: $CdCO_3$, $MnCO_3$, $Zn_5(OH)_6(CO_3)_2$, $Co(OH)_2$, and $Ni(OH)_2$. This fact indicates that the concentration of heavy metals in the solution can be regulated by the solubility of dispersed carbonates.

The comparison of the experimental Cu, Pb, and Zn contents in slightly saline extracts from the soils studied with the theoretical solubilities of their compounds derived from solubility diagrams shows that Cu, Pb, and Zn hydroxides and carbonates can be formed in ordinary chernozems under usual neutral conditions (Minkina et al., 2005). The formation of metal carbonates at pH 7–8 was also supposed in other soils (Fassbender, Seekamp, 1977; Chemistry of Heavy Metals, Arsenic, and Molybdenum in Soils, 1985; Lead in the Environment, 1987; Gorbatov, 1988; Ponizovskii, Mironenko, 2001; Okorkov, Mazirov, 2003; Han, 2007). The binding strength of metal ions to carbonate particles depends on the size of these particles (Mineev et al., 1984).

Heavy metal ions weakly bound to carbonates can be extracted from soils by acid reagents or a 1 M CH_3COONa solution with pH 5 by the methods of Tessier (1979) and Shuman (1979). However, acid reagents extracts additional amounts of metal specifically sorbed on Fe and Mn (hydr)oxides.

The solubility of carbonates and the release of metals from them depend on the degree of dispersion of carbonates. Finely dispersed (active) carbonates (particle size <20 nm) retain metal ions in the weakly bound state. They presumably can be extracted from soils by a 1 N AAB solution with pH 4.8. This is confirmed by the agreement between the contents of carbonates dissolved by this reagent and their active forms in the soil determined in a 0.2 N $(NH_4)_2C_2O_4$ solution by the Drouineau–Galet method (Peterburgskii, 1968). Their portion is usually no more than 10–15% of the total carbonate content. Similar shares of active carbonates were found in ordinary chernozems of the Myasnikovskii region, Rostov oblast, by Ageev et al. (1996). The authors noted the similar seasonal dynamics of active and total carbonates. This group of carbonates largely supplies the clean and contaminated calcareous soils with mobile metals (especially zinc) (Mineev et al., 1984).

In calcareous soils, metals that are retained on the surface of carbonates by chemisorption or form their own minerals (carbonate, hydroxycarbonates, metal oxides) occur in the strongly bound form. According to the classification suggested by Zyrin et al. (1974), the chemisorbed metals and the metals forming difficultly soluble compounds (SP < 10^{-6}) are referred to as fixed metals. A difficulty in the determination of chemisorbed heavy metal compounds is that these compounds are transferred into solution upon dissolution of carbonates.

The determination of the own minerals of heavy metals is an even more difficult problem. These minerals cannot be detected in soil by extraction methods. They can have a very complex composition due to coprecipitation of several compounds that can also exist individually, which makes unfeasible their identification (Santillan-Medrano, Jurinak, 1975).

None of fractionation schemes involves the separation of this metal group, because their minerals occur in small amounts and cannot be selectively extracted. It is suggested that metals usually form no separate phase in the soil but are retained by soil components. At the same time, some minerals of heavy metals, whose presence in soil was proved by EXAFS spectroscopy, are given in Table 2.4.

Reagents used for extracting heavy-metal ions from carbonates only partly extract difficultly soluble metal compounds. It was experimentally shown that an ordinary chernozem sample contaminated with metals retained a part of carbonates after the treatment with a 1 N $NaCH_3COO$ with pH 5, in distinction from their complete dissolution in a clean soil. The share of undissolved carbonates increased with increasing amount of $CaCO_3$ in the soil. It is possible that the metals remained in this fraction occur in the form of difficultly soluble carbonates and hydroxides.

(3) *Fractionation of metal compounds bound to nonsilicate Fe, Al, and Mn compounds.*

The content of a metal specifically sorbed by Fe, Al and *Mn* (hydr)oxides can be found by subtracting its content retained by the SEC (the content of metals in 1 N AAB) and carbonates (the content of metals in 1 N $NaCH_3COO$) from the content of the metal in 1 N HCl.

The content of a metal strongly retained by Fe, Al, and Mn (hydr)oxides can be found as the difference between its total content retained by nonsilicate minerals (in the Tessier fractionation scheme, metals in 0.04 M $NH_2OH \cdot HCl$) and the content of metal specifically sorbed by finely dispersed nonsilicate minerals.

(4) *Fractionation of metal compounds bound to organic matter.* As was substantiated in Section 2.2, the increase in the content of metals extracted by the mixed reagent (1 N AAB + 1% EDTA), compared to their content in 1 N AAB, can be attributed to the metals weakly bound to organic and organomineral substances into outer-sphere surface complexes.

The extraction of soil with 1% EDTA transfers into the solution only unstable organic and organomineral compounds of heavy metals. The treatment with 30% H_2O_2 results in the almost complete oxidation of organic substances with the release of bound metals. The latter approach (acid extraction from soils before and after their treatment with H_2O_2) was repeatedly used (Zyrin et al., 1974; Tessier et al., 1979) for the determination of total metals retained by soil organic matter. Therefore, the difference between the total content of metals after the treatment of soil with 30% H_2O_2 and their amount extracted with 1% EDTA can be indicative of the amount of metals strongly bound to organic and organomineral soil substances; they mainly consist of stable inner-sphere complexes (chelates).

A similar approach to the separation of metals weakly and strongly bound to organic matter was used for the sequential extraction of metals with 0.1 M EDTA and 0.1 M $K_4P_2O_7$ + 0.1 M NaOH (Ladonin, Plyaskina, 2003).

Thus, the combined fractionation scheme presumably separates all heavy metal ions bonded to organic matter and Fe, Al, and Mn (hydr)oxides into the groups of weakly and strongly bound metals. This enables us to assess the participation of these soil components in the fixation of heavy metals, which is of theoretical and practical importance for estimating and predicting the ecological state of contaminated soils.

Chapter 3

COMPOSITION OF COPPER, ZINC, AND LEAD COMPOUNDS IN CLEAN AND CONTAMINATED SOILS OF THE LOWER DON BASIN (SOUTHERN FEDERAL DISTRICT OF RUSSIAN FEDERATION)

The effect of contaminated soils on the ecological status of an ecosystem directly depends on the group composition of metal compounds. Mobile metal compounds in soil solid phases are in equilibrium with the liquid phases. The equilibrium between metal compounds in a heterogeneous soil system is dynamic, which creates conditions for their transformation.

The transformation of metal compounds in contaminated soils depend on the nature of metal, its input into the soil, the presence of attendant metals, the properties of contaminated soils, and the residence time of metals in the soils (Fig. 2.1). To elucidate these problems, the group composition of heavy metal compounds in soils of the Lower Don basin and its changes under the effect of anthropogenic factors were studied.

The Lower Don basin (in the Rostov oblast, Russian Federation) is located in two soil zones. These are steppes with ordinary and southern chernozems and dry steppes with chestnut soils. The steppe zone occupies 6.2 million ha (two thirds of the oblast area) in the most humid regions of the oblast. The dry-steppe zone occupies 3.4 million ha in the eastern part of the oblast.

Meadow-chernozemic soils (semihydromorphic analogues of chernozems) are also found. They occupy only 2% of the total area in Rostov oblast. These soils develop under increased moistening conditions due to relatively shallow groundwater level (3–6 m).

Alluvial meadow soils occupy 1.5% of the total area. They develop under a permanent or transient effect of alluvial conditions in river floodplains at a groundwater table depth of less than 2 m (Val'kov et al., 2002).

The natural conditions of the chernozemic zone are most favorable for agriculture. Soils were developed under continental climatic conditions with an annual precipitation of 400–500 mm. Calcareous loess-like clays and loams are the predominant parent rocks of chernozems.

The zone of chestnut soils is characterized by an arid climate with an annual precipitation of 300–350 mm. Carbonate and carbonate-sulfate loess-like loams and clays are the main soil-forming rocks of chestnut soils.

Lands in the Lower Don basin have been intensively used over centuries; there are almost no soils free from anthropogenic impacts.

3.1. Heavy Metal Compounds in Clean Soils

Thick calcareous low-humus clay loamy ordinary chernozem on loess-like loams in the steppe zone of Rostov oblast (Oktyabr'skii region) and medium-thick clay loamy chestnut soil on loess-like loams in the dry-steppe zone of Rostov oblast (Zimovniki region) were studied. Some properties of soils are given in Table 3.1. The upper (0- to 20-cm) layers of these soils were studied, because the soil-forming factors and soil properties developed in the steppe and dry-steppe zone of the Lower Don basin favor a weak migration of exogenic heavy metal compounds.

The total metal content in the 0- to 20-cm layer of the original soils is 42–44 mg/kg for Cu, 20–25 mg/kg for Pb, and 65–69 mg/kg for Zn (Table 3.2). These parameters correspond to the background contents of metals in ordinary chernozems and chestnut soils (Zakrutkin, Shishkina, 1997; Zakrutkin, Shkafenko, 1997; Nikityuk, 1998; Samokhin, 2003). They exceed the clarke values for Cu and Pb by more than 2 times and that for Zn by 1.3 times (According to A. P. Vinogradov (1957), the metal clarkes for soils are as follows (mg/kg): Cu, 20; Pb, 10; Zn, 50).

Table 3.1. Chemical and physical properties of the soils studied (0 to 20-cm layer)

Soil	Humus, %	CaCO$_3$, %	pH	N-NO$_3$, mg/100 g	P$_2$O$_5$ mob., mg/100 g	K$_2$O exch., mg/100 g	Exchangeable ions, meq/100 g Ca^{2+}	Mg^{2+}	Na$^+$	<0.01 mm, %	<0.001 mm, %
Calcareous ordinary chernozem	3,9	0,4	7,5	0,8	3,2	24,8	29,5	5,5	0,1	53,1	32,4
Solonetzic chestnut soil	2,6	0,1	7,8	0,6	1,2	38,0	20,2	4,5	2,4	47,7	29,5

Table 3.2. Group composition and mobility (Km) parameters of Cu, Pb and Zn compounds in clean soils

Element	Total content, mg/kg	WB/SB*	Km	Weakly bound compounds: exchangeable/complex/ specifically sorbed
Ordinary chernozem				
Cu	44	5/95	0.1	13/9/78
Pb	25	12/88	0.1	18/9/73
Zn	69	10/90	0.1	7/1/92
Chestnut soil				
Cu	42	7/93	0.1	10/7/83
Pb	20	15/85	0.2	19/11/70
Zn	65	12/88	0.1	5/4/92

* Weakly/strongly bound compounds, % of the total content.
** % of weakly bound compounds.

However, there is no ecological hazard for the regional soils, because the major part of metals in the soil (88–95% of their total content) occurs in the strongly bound state (Table 3.2). This is a reason for the insufficient supply of plants with zinc and especially copper, because weakly bound compounds make up 5–15% of the total metal contents in ordinary chernozem and chestnut soil. The composition of weakly bound metal compounds of the highest ecological significance in soils is shown in Table 3.3.

Weakly bound compounds of the metals studied mainly contain specifically sorbed forms (70–92% of the total amount of weakly bound metal compounds).

Their highest share (92%) is found in the weakly bound zinc compounds. The content of exchangeable and complex metal compounds is insignificant.

The contents of exchangeable zinc in chernozem and chestnut soil are slightly higher compared to other metals, in spite of the slightly alkaline reaction of these soils. Exchangeable lead forms make up almost one fifth of weakly bound compounds in soils. The size of ions can have an important effect in exchange reactions (Plekhanova et al., 2001). The Pb^{2+} and Ca^{2+} ions have similar radii (1.3–1.2 and 1.1–1.1 Å, respectively), and Ca^{2+} ions are predominant in the exchange complex of the soils studied.

The role of separate soil components in the retention of metals was established using a combined scheme of metal compounds (Table 3.3).

Table 3.3. Group composition of Cu, Pb and Zn in clean chernozem (combined scheme of fractionation)

Element	Weakly bound compounds (WB)			Strongly bound compounds (SB)				Sum of fractions
	Exchangeable AAB/MgCl$_2$	Complex	Specifically sorbed		With organic matter	With Fe and Mn (hydr)oxides	With silicates	
			on carbonates	on Fe and Mn (hydr)oxides				
mg/kg								
Cu	0,3±0,1/ 0,3±0,1	0,2±0,06	1,7±0,4	0,2±0,05	4,2±1,0	0,9±0,1	36,9±5,2	44,4±5,9
Pb	0,6±0,2/ 0,4±0,1	0,3±0,1	1,6±0,6	0,7±0,2	6,5±1,1	1,8±0,08	14,3±2,1	25,6±4,7
Zn	0,4±0,1/ 0,3±0,1	0,3±0,1	6,3±1,8	0,3±0,1	1,0±0,3	6,2±1,6	55,9±4,4	70,3±7,0
Relative contents of weakly and strongly bound HM compounds, % of their group composition								WB/SB
Cu	13	8	71	8	10	2	88	5/95
Pb	13	11	53	23	29	8	63	12/88
Zn	4	4	88	4	2	10	88	10/90

The predominance of strongly bound compounds in the soils studied is mainly ensured by the fixation of metals in lattices of silicate minerals (54–82% of the total amount and 63–88% of the strongly bound compounds). This is due to the mineralogy of soil-forming rocks in Rostov oblast. Yellow-brown loess-like loams and clays of Ciscaucasian plains partly inherited the metal-rich stable minerals from the original rocks. Some metal ions released during weathering and pedogenesis are strongly fixed in the structure of clay minerals. Trioctahedral montmorillonite, vermiculite, chlorite, and biotite have the highest capacities of strongly fixing metals (Motuzova, 1988; Dobrovol'skii, 1997; Ladonin, 1997).

The mobility of Cu, Pb, and Zn in the soils studied is mainly due to their compounds retained by carbonates (53–88% of the weakly bound compounds). The highest affinity to carbonates is observed for zinc. The share of metal forms specifically sorbed by them in the weakly bound compounds of background soils is almost 90%.

This metal fraction was not identified earlier in typically chernozem (Pampura et al., 1993). The presence of heavy metal compounds presumably bound to carbonates is a characteristic feature of ordinary chernozems related to the specific micellar forms of carbonates in these soils. These carbonate structures have a high specific surface area favoring an increase in their dissolution rate and activity in the solution.

This ratio of metal compounds in the soils studied was developed under the effect of zonal soil-forming factors and the long period of agricultural use.

3.2. HEAVY METAL COMPOUNDS IN CONTAMINATED SOILS

The system of soil heavy metal compounds, as well as any natural system, tends to a more stable state. In the soils studied, this corresponds to the predominance of strongly bound metal forms. In contaminated soils, the development of similar ratios between the different groups of metal compounds is hampered by the formation of additional amounts of mobile metal forms.

3.2.1. Transformation of Copper, Zinc, and Lead Compounds under Soil Contamination in a Pot Experiment

Under conditions of a pot experiment, the soils under study were contaminated with Cu, Pb, and Zn salts. The contents of metals in the soils were determined one and two years after contamination. The comparative analysis of metal compounds in the contaminated and clean soils revealed the effect of metal properties, application rates, time of residence, and soil properties.

3.2.1.1. Procedure of Pot Experiments

The effect of different rates of Cu, Zn, and Pb applied separately and simultaneously on the sorption of metals by soils, their distribution among different soil compounds, and translocation to plants was studied in a pot experiment (Table 3.4). Samples of ordinary chernozem and chestnut soil, whose properties were described above (Table 3.1), were used in the experiment.

Table 3.4. Addition dozes of Cu, Zn and Pb (mg/kg) in ordinary chernozem and chestnut soil

Separate addition			Joint addition of Cu, Pb and Zn
Cu	Pb	Zn	
3	6	23	Cu 3 + Pb 6 + Zn 23
10	25	50	Cu 10 + Pb 25 + Zn 50
30	32	75	Cu 55 + Pb 32 + Zn 100
55	55	100	Cu 100 + Pb 100 + Zn 300
100	100	300	

The selected application rates of metals should ensure metal contents close to 2–3 MCL, which correspond to the contamination level of soil in Rostov oblast with these metals (Ecological Atlas …, 2000).

Draining material (a 3-cm thick layer of claydite and a 3-cm thick layer of washed river sand) was placed in polyethylene bags of 4 l, which were used as vegetation pots. The layer was covered with 4 kg of soil sieved through a 5-mm sieve and mixed with dry heavy-metal salts in accordance with the experimental design (Table 3.4). The samples were wetted to field capacity.

Experiments were conducted in triplicate. Metals were applied as Cu, Pb, and Zn acetates. Acetates were used for simulating the contamination of soils because of their good solubility and, hence, rapid and complete interaction with the soil material. In distinction from mineral acid soils, the hydrolysis of heavy metal acetates is not accompanied by an abrupt shift toward a strongly acid reaction; acetate anions are natural products of plant metabolism and cannot significantly affect the nutritive regime of soil.

After soil composting for a month, a test crop was planted. The Odesskii-100 cultivar of barley (*Hordeum sativum distichum*) was used. Seeds were planted to a depth of 4 cm at a rate of 20 seeds per pot with account for the feeding area in accordance with the common row cropping method.

Soils were maintained at a level of 60% of maximum moisture capacity during the vegetation period. Barley was harvested at the dough stage. Soil samples were taken before the beginning of the experiment and during two years after harvesting.

3.2.1.2. Results of Pot Experiment

After the artificial contamination of ordinary chernozem and chestnut soil in the studied range of metal concentrations, their total content in the soil increased by 1.2–5.3 times (Tables 3.5–3.7). The absolute content of three mobile metal fractions increased to a higher extent. Differences between the treatments were statistically significant.

The group composition of Cu, Pb, and Zn compounds in the contaminated soils revealed tendencies in the transformation of their compounds. The contamination disturbed the natural proportions of metal compounds in these soils (Tables 3.8, 3.9).

The contamination of soils entailed a decrease in the content of strongly bound metal compounds from the manifest predominance (their share in the clean soils was 88–95% of the total metal content) to a level of 55% of the total amount. The mobility coefficients of metals, which indicated the ratio between the weakly and strongly bound compounds, increased by 4–7 times at the maximum level of contamination. The main tendency in the occurring changes was an increase in the share of more mobile compounds with increasing rate of the metal applied. This process was enhanced at the simultaneous application of metals to the soil (Table 3.10).

Table 3.5. The total content and weakly bound compounds of Cu, Zn and Pb in the case of monoelement contamination of the chernozem, mg/kg (n=9)

Dose of metal, mg/kg	Exchangeable compounds 1 year	Exchangeable compounds 2 years	Complex compounds 1 year	Complex compounds 2 years	Specifically sorbed compounds 1 year	Specifically sorbed compounds 2 years	Total content 1 year	Total content 2 years
Cu								
Control (no added metal)	0,3	0,3	0,2	0,2	1,8	1,9	44	44
3	0,4	0,4	0,6	0,8	3,3	3,2	46	44
10	0,8	0,5	0,5	1,0	4,2	4,6	53	52
30	1,1	0,7	1,0	2,0	10,6	10,8	73	71
55	1,3	1,9	1,6	3,1	17,0	18,6	100	93
100	4,4	3,1	6,7	10,8	25,1	25,0	135	139
LSD$_{0,95}$	0,9	0,8	1,4	2,3	4,5	5,8	15	14
Pb								
Control (no added metal)	0,6	0,6	0,3	0,3	2,4	2,3	25	24
6	0,7	0,6	0,6	1,0	2,2	2,3	33	28
25	0,8	0,8	0,7	1,7	3,6	2,3	42	44
32	1,4	1,1	0,9	3,1	4,8	3,4	60	56
55	3,4	2,2	2,1	4,4	5,9	5,6	78	76
100	8,2	4,5	5,8	9,4	21,7	19,1	127	125
LSD$_{0,95}$	1,2	0,8	1,6	2,2	3,3	3,7	13	14
Zn								
Control (no added metal)	0,5	0,4	0,1	0,3	6,5	6,6	69	65
23	1,1	0,7	2,0	2,3	16,3	14,8	93	89
50	2,5	1,5	1,4	1,9	31,9	46,2	119	121
75	4,1	3,6	0,6	0,6	48,4	60,6	140	137
100	7,5	4,0	6,6	4,0	54,5	67,2	165	159
300	25,4	22,5	29,4	20,7	57,6	90,7	365	360
LSD$_{0,95}$	1,5	1,9	3,5	3,5	5,6	6,7	19	20

Table 3.6. The total content and weakly bound compounds of Cu, Zn and Pb in the case of monoelement contamination of the chestnut soil, mg/kg (n = 9)

Dose of metal, mg/kg	Exchangeable compounds 1 year	Exchangeable compounds 2 years	Complex compounds 1 year	Complex compounds 2 years	Specifically sorbed compounds 1 year	Specifically sorbed compounds 2 years	Total content 1 year	Total content 2 years
Cu								
Control (no added metal)	0,3	0,2	0,2	0,3	2,4	2,3	42	39
3	0,5	0,4	0,3	0,5	2,6	4,2	42	40
10	0,9	0,8	0,1	0,5	4,5	5,6	47	46
30	1,1	2,6	1,3	0,6	8,5	12,8	64	65
55	3,1	2,3	1,2	3,4	14,4	14,2	92	86
100	7,9	4,9	4,3	7,6	22,3	23,1	125	129
LSD$_{0,95}$	0,7	1,1	1,2	1,8	5,4	4,8	13	15
Pb								
Control (no added metal)	0,5	0,4	0,3	0,4	1,9	2,0	20	22
6	0,4	0,7	0,4	1,3	2,7	3,0	28	28
20	1,3	1,0	0,1	1,2	2,6	2,5	32	39
32	1,8	1,7	0,3	2,1	4,5	3,8	53	53
55	4,6	3,4	0,8	5,7	8,1	6,4	71	75
100	10,8	11,3	3,6	11,5	21,9	17,6	119	115
LSD$_{0,95}$	0,7	1,4	1,7	2,7	4,3	3,3	11	13
Zn								
Control (no added metal)	0,4	0,4	0,3	0,4	7,7	8,6	65	59
23	1,9	1,2	0,6	0,8	14,5	15,4	83	78
50	2,9	2,1	0,5	0,6	33,1	38,2	109	100
75	4,4	3,9	0,2	0,1	47,0	71,2	139	135
100	10,8	6,7	5,3	3,6	57,1	74,6	162	165
300	35,1	20,6	21,8	8,9	53,2	90,7	355	356
LSD$_{0,95}$	2,1	1,5	2,9	4,6	6,6	5,5	16	19

Table 3.7. The total content and weakly bound compounds of Cu, Zn and Pb in the case of polyelement contamination of an ordinary chernozem, mg/kg (n = 9)

Dose of metal, mg/kg	Exchangeable compounds 1 year	Exchangeable compounds 2 years	Complex compounds 1 year	Complex compounds 2 years	Specifically sorbed compounds 1 year	Specifically sorbed compounds 2 years	Total content 1 year	Total content 2 years
Cu								
Control (no added metal)	0,3	0,3	0,2	0,2	1,8	1,9	44	44
Cu 3 + Pb 6 + Zn 23	0,7	0,6	0,4	0,7	3,7	4,1	45	43
Cu 10 + Pb25+ Zn 50	1,4	1,0	0,6	1,5	5,8	7,5	54	53
Cu55+ Pb 32 + Zn 100	2,1	2,7	1,5	3,0	22,7	24,2	101	98
Cu100+Pb100+Zn 300	6,6	5,6	6,1	11,0	32,3	32,1	143	150
LSD$_{0,95}$	1,1	1,2	2,0	2,5	6,5	5,3	14	14
Pb								
Control (no added metal)	0,6	0,6	0,3	0,2	2,4	2,3	25	24
Cu 3+ Pb 6 + Zn 23	0,6	0,6	1,3	2,0	2,9	2,8	30	25
Cu 10 + Pb25 + Zn 50	1,0	0,7	1,2	2,2	4,7	5,9	45	46
Cu 55+Pb 32 + Zn 100	3,3	2,2	0,9	4,7	9,5	9,4	67	64
Cu100+Pb100+Zn 300	7,5	6,8	3,8	13,0	22,7	17,4	126	117
LSD$_{0,95}$	1,5	1,4	2,0	2,7	2,7	4,5	13	14
Zn								
Control (no added metal)	0,5	0,4	0,1	0,3	6,5	6,6	67,0	65,0
Cu 3 + Pb 6 + Zn 23	1,9	1,1	3,0	3,5	16,5	19,5	90,4	82,4
Cu 10 + Pb25+ Zn 50	5,9	4,2	0,5	1,6	41,3	45,9	115,7	110,0
Cu55+ Pb 32 + Zn 100	12,9	8,5	4,3	1,4	62,9	76,5	165,0	158,0
Cu100+Pb100+Zn 300	24,3	16,9	36,1	30,6	59,3	96,6	370,0	376,0
LSD$_{0,95}$	1,4	1,4	2,8	2,5	6,3	6,7	15,9	185,2

Table 3.8. Group composition and mobility parameters (Km) of Cu, Pb, and Zn compounds in the case of monoelement contamination of the chernozem (parallel extraction)

Dose of metal, mg/kg	WB/SB *		Exchangeable/ complex/ specifically sorbed **		K_m	
	1 year	1 year	1 year	2 year	1 year	2 year
Cu						
Control (no added metal)	5/95	5/95	13/9/78	13/8/79	0,1	0,1
3	9/91	9/91	9/14/77	9/18/73	0,1	0,1
10	11/89	12/88	15/9/76	8/16/76	0,1	0,1
30	18/82	20/80	9/8/83	5/15/80	0,2	0,3
55	20/80	26/74	7/8/85	8/13/79	0,3	0,4
100	27/73	28/72	12/19/69	8/28/64	0,4	0,4
Pb						
Control (no added metal)	12/88	14/86	18/9/73	19/9/72	0,1	0,2
6	12/88	14/86	20/17/63	15/26/59	0,1	0,2
25	12/88	11/89	16/14/70	17/35/48	0,1	0,1
32	12/88	14/86	20/13/67	14/41/45	0,1	0,2
55	14/86	16/84	30/18/52	18/36/46	0,2	0,2
100	28/72	26/74	23/16/61	14/29/57	0,4	0,4
Zn						
Control (no added metal)	10/90	11/89	7/1/92	5/4/91	0,1	0,1
Dose of metal, mg/kg	WB/SB *		Exchangeable/ complex/ specifically sorbed **		K_m	
	1 year	1 year	1 year	2 year	1 year	2 year
23	20/80	20/80	6/10/84	4/13/83	0,3	0,3
50	30/70	41/59	7/4/89	3/4/93	0,4	0,7
75	38/62	48/52	8/1/91	6/1/93	0,6	0,9
100	42/58	47/53	11/10/79	5/5/90	0,7	0,9
300	31/69	37/63	23/26/51	17/15/68	0,5	0,6

* The WB/SB values are given in percent of the total content; ** the contents of the exchangeable/complex/specifically sorbed metals are given in percent of the content of weakly bound metals.

Table 3.9. Group composition and mobility parameters (Km) of Cu, Pb, and Zn compounds in the case of monoelement contamination of the chestnut soil (Parallel extraction)

Dose of metal, mg/kg	WB/SB *		Exchangeable/ complex/ specifically sorbed **		K_m	
	1 year	1 year	1 year	2 year	1 year	2 year
Cu						
Control (no added metal)	7/93	8/92	10/7/83	7/11/82	0,1	0,1
3	7/93	13/87	15/9/76	8/10/82	0,1	0,2
10	13/87	15/85	16/2/82	12/7/81	0,2	0,2
30	17/83	25/75	10/12/78	16/4/80	0,2	0,3
55	21/79	23/77	17/6/77	12/17/71	0,3	0,3
100	28/72	28/72	23/12/65	14/21/65	0,4	0,4
Pb						
Control (no added metal)	15/85	14/86	19/11/70	14/14/72	0,2	0,2
6	14/86	18/82	11/11/78	14/26/60	0,1	0,2
25	13/87	13/87	32/3/65	21/26/53	0,2	0,2
32	13/87	15/85	27/5/68	22/28/50	0,2	0,2
55	20/80	21/79	34/6/60	22/37/41	0,3	0,3
100	30/70	35/65	30/10/60	28/28/44	0,4	0,5
Zn						
Control (no added metal)	12/88	15/85	5/4/92	4/4/92	0,1	0,2
23	21/79	22/78	11/4/85	7/5/88	0,3	0,3
50	34/66	41/59	8/1/91	5/1/94	0,5	0,7
75	37/63	56/44	8/1/91	5/1/94	0,6	1,3
100	45/55	52/48	15/7/78	8/4/88	0,8	1,1
300	31/69	34/66	32/20/48	17/8/75	0,5	0,5

* The WB/SB values are given in percent of the total content; ** the contents of the exchangeable/complex/specifically sorbed metals are given in percent of the content of weakly bound metals.

Table 3.10. Group composition and mobility parameters (Km) of Cu, Pb, and Zn compounds in the case of polyelement contamination of an ordinary chernozem (Parallel extraction)

Dose of metal, mg/kg	WB/SB *		Exchangeable/ complex/ specifically sorbed **		K_m	
	1 year	1 year	1 year	2 year	1 year	2 year
Cu						
Без внесения	5/95	5/95	13/9/78	13/8/79	0,1	0,1
Cu 3 + Pb 6 + Zn 23	11/89	12/88	15/8/77	11/13/76	0,1	0,1
Cu 10 + Pb25+ Zn 50	15/85	19/81	18/8/74	10/15/75	0,2	0,2
Cu 55 + Pb 32 + Zn 100	26/74	31/69	8/6/86	9/10/81	0,4	0,5
Cu 100 + Pb 100 + Zn 300	32/68	33/67	15/13/72	11/23/66	0,5	0,5
Pb						
Без внесения	12/88	14/86	18/9/73	19/9/72	0,1	0,2
Cu 3 + Pb 6 + Zn 23	17/83	20/80	13/27/60	11/37/52	0,2	0,3
Cu 10 + Pb25+ Zn 50	16/84	20/80	14/17/69	8/25/67	0,2	0,3
Cu 55 + Pb 32 + Zn 100	21/79	25/75	24/7/69	13/29/58	0,3	0,3
Cu 100 + Pb 100 + Zn 300	27/73	32/68	22/11/67	18/35/47	0,4	0,5
Zn						
Без внесения	10/90	11/89	7/1/92	5/4/91	0,1	0,1
Cu 3 + Pb 6 + Zn 23	23/77	29/71	9/14/77	4/15/81	0,3	0,4
Cu 10 + Pb25+ Zn 50	41/59	47/53	12/1/87	8/3/89	0,7	0,9
Cu 55 + Pb 32 + Zn 100	49/51	54/46	16/5/79	10/2/88	1,0	1,2
Cu 100 + Pb 100 + Zn 300	32/68	38/62	20/30/50	12/21/67	0,5	0,6

* The WB/SB values are given in percent of the total content; ** the contents of the exchangeable/complex/specifically sorbed metals are given in percent of the content of weakly bound metals.

A higher activity of exchange processes in the transformation of heavy metal compounds in all experimental treatments was observed for the contaminated chestnut soil compared to chernozem. Along with this general tendency, each metal exhibited specific features depending on the properties of its atoms and the soil. These tendencies could be revealed from the comparison of results in absolute and relative values (portions of different metal fractions in their total content in the soil and in the group of compounds to which the fractions belong).

The combined fractionation scheme revealed the participation of soil components in the strong and weak fixation of metals.

GENERAL CHANGES IN THE GROUP COMPOSITION OF CU, PB, AND ZN COMPOUNDS IN CHERNOZEM AND CHESTNUT SOIL UNDER CONTAMINATION

Contamination significantly increased the total content of metals, especially in weakly bound compounds (Table 3.11).

Qualitative changes in metal mobility also occurred, which were confirmed by changes in the proportions of different fractions in the group of weakly bound metal compounds. The mobility of metals increased predominantly due to mobile metal compounds bound to organic substances, the relative contents of which increased by 3–5 times (Tables 3.12, 3.13). This decreased the participation of carbonate-bound fractions of weakly bound metals.

An increase in the absolute content of heavy metals in the residual fraction was observed. At the same time, the relative content of this fraction significantly decreased (by 2–4 times) with increasing soil contamination (Tables 3.12, 3.13), which was indicative of a limited resistance of soils to contamination. It was proposed to use the share of metal in the crystal lattices of primary and secondary minerals in its total content in the soil as the recalcitrant factor (RF) of soils (Knox et al., 2000).

The contamination of chernozems with metals did not affect the predominance of their compounds strongly bound by soil components. However, their contribution to the increase in the total metal content decreased by 13–30% because of the active involvement of the metals added into weakly bound compounds. The higher was the load, the more intensive was this process.

Table 3.11. The total content and group composition of Cu, Pb and Zn in polluted chernozem, mg/kg

Dose of metal, mg/kg	Weakly bound compounds (WB)			Strongly bound compounds (SB)				Sum of fractions
	Exchangeable AAB/MgCl$_2$	Complex	Specifically sorbed		With organic matter	With Fe and Mn (hydr)oxides	With silicates	
			on carbonates	on Fe and Mn (hydr)oxides				
Cu								
Control (no added metal)	0,3±0,1/ 0,3±0,1	0,2±0,06	1,7±0,4	0,2±0,05	4,2±1,0	0,9±0,1	36,9±5,2	44,4±5,9
100	3,1±0,8/ 2,1±0,7	10,8±2,3	19,7±4,7	5,3±1,5	44,9±6,3	20,0±3,3	40,4±3,8	143,2±13,7
300	14,0±3,9/ 7,9±2,0	44,9±4,2	36,5±4,9	18,6±5,4	106,5±10,2	76,1±7,8	69,2±4,8	359,7±18,8
Pb								
Control (no added metal)	0,6±0,1/ 0,4±0,1	0,3±0,1	1,6±0,5	0,7±0,2	6,5±1,1	1,8±0,08	14,3±2,1	25,6±4,7
100	4,5±0,8/ 3,2±0,7	9,4±1,9	14,3±2,9	4,8±1,8	41,0±6,4	27,4±6,4	20,9±2,7	121,0±12,9
300	12,4±1,5/ 10,2±2,5	46,0±5,0	32,8±4,0	12,0±3,5	114,2±12,7	74,9±7,9	36,9±4,3	327,0±19,0
Zn								
Control (no added metal)	0,4±0,1/ 0,3±0,04	0,3±0,04	6,3±1,3	0,3±0,1	1,0±0,3	6,2±1,6	55,9±4,4	70,3±7,0
100	4,0±0,8/ 2,2±0,4	4,0±1,2	50,5±3,7	16,7±4,9	2,4±0,9	45,4±6,6	65,1±4,4	186,3±12,8
300	22,5±3,7/ 15,2±3,0	20,7±3,6	50,7±5,2	40,0±5,6	3,9±1,1	104,5±11,3	84,0±5,7	319,0±18,0

Table 3.12. Relative contents of Cu, Pb and Zn compounds in polluted chernozem, % of the total content

Dose of metal, mg/kg	Weakly bound compounds (WB)			Strongly bound compounds (SB)				WB/SB
	Exchangeable, MgCl$_2$	Complex	Specifically sorbed		With organic matter	With Fe and Mn (hydr)oxides	With silicates	
			on carbonates	on Fe and Mn (hydr)oxides				
Cu								
Control (no added metal)	0,7	0,5	3,8	0,5	9,5	2,0	83,1	5/95
100	1,5	7,5	13,8	3,7	31,4	14,0	28,2	26/74
300	2,2	12,5	10,1	5,2	29,6	21,2	19,2	30/70
Pb								
Control (no added metal)	1,6	1,2	6,3	2,7	25,4	7,0	55,9	12/88
100	2,6	7,8	11,8	4,0	33,9	22,6	17,3	25/75
300	3,1	14,1	10,0	3,7	34,9	22,9	11,3	31/69
Zn								
Control (no added metal)	0,4	0,4	9,0	0,4	1,4	8,8	79,5	10/90
100	1,2	2,1	27,1	9,0	1,3	24,4	34,9	39/61
300	4,8	6,5	15,9	12,5	1,2	32,8	26,3	40/60

Table 3.13. Relative contents of Cu, Pb and Zn forms in groups of weakly and strongly bound compounds in polluted chernozem

Dose of metal, mg/kg	Weakly bound compounds (WB)		Specifically sorbed		Strongly bound compounds (SB)		
	Exchangeable, MgCl$_2$	Complex	on carbonates	on Fe and Mn (hydr)oxides	With organic matter	With Fe and Mn (hydr)oxides	With silicates
Cu							
Control (no added metal)	13	8	71	8	10	2	88
100	6	28	52	14	43	19	38
300	7	42	34	17	42	31	27
Pb							
Control (no added metal)	13	11	53	23	29	8	63
100	10	30	45	15	46	31	23
300	10	46	32	12	51	33	16
Zn							
Control (no added metal)	4	4	88	4	2	10	88
100	3	5	69	23	2	40	58
300	12	16	40	32	2	54	44

Iron and manganese oxides actively participated in the strong fixation of metals in the contaminated soils (Tables 3.13, 3.14).

Table 3.14. The ratio of weakly bound /strongly bound compounds of Cu, Pb and Zn with organic matter and Fe and Mn (hydr)oxides

Dose of metal, mg/kg	The ratio of HM concentration (mg/kg) with		The ratio of HM relative concentration, % of their total contents within	
	organic matter	Fe and Mn (hydr)oxides	organic matter	Fe and Mn (hydr)oxides
Cu				
Control (no added metal)	0,2/4,2	0,2/0,9	5/95	18/82
100	11/45	5/20	20/80	20/80
300	45/106	19/76	30/70	20/80
Pb				
Control (no added metal)	0,3/6,5	0,7/1,8	4/96	28/72
100	9/41	5/27	18/82	16/84
300	46/114	12/75	29/71	14/86
Zn				
Control (no added metal)	0,3/1	0,3/6	23/77	5/95
100	4/2	17/45	67/33	27/73
300	21/4	40/104	84/16	28/72

The capacity of strongly fixing heavy metals by crystallized particles of Fe and Mn oxides and hydroxides should be considered a general tendency for soils of different genesis (Pinsky, 1992). It is suggested that heavy metal ions are fixed in the nonexchangeable form by chemisorption in the result of substitution of divalent metal cations for two H^+ ions of OH or OH_2 groups on the surface of oxides. The affinity mainly depends on the hydrolyzability and electronegativity of metal cations (McBride, 1989); therefore, readily hydrolysable metals (Cu, Pb, Zn) have the highest ability to be strongly fixed on the surface of mineral particles by the exchange for hydroxo groups. No less than half the total content of metals is bound to iron hydroxides in different soils (Dobrovol'skii, 1997). These soils are classified among the soils with the leading effect of pedogenic factors on the fractional composition of microelement compounds (Motuzova, Aptikaev, 2006).

In the treatments with different technogenic loads (100 and 300 mg/kg), changes in the group composition of heavy metals were similar in character but their degree of manifestation increased with increasing contamination.

All soil components were involved in reactions with metals, but their contributions to the fate of separate elements were different.

FORMATION OF THE GROUP COMPOSITION OF METAL COMPOUNDS IN CONTAMINATED SOILS

Copper Compounds

Group of weakly bound copper compounds. In clean chernozem, most copper was strongly fixed by soil components and only 5% occurred in the potentially mobile, weakly bound state (Table 3.12). Copper ions arrived in chernozem primarily increased the content of specifically sorbed compounds (their share was 78–79%), as well as exchangeable and complex forms. The ratio between these fractions in the contaminated soils changed in comparison to the clean soils because of differences in their formation rates. A dynamic formation–transformation equilibrium was established in different groups of mobile metal compounds and followed by the accumulation of relatively more stable forms.

The group of weakly bound copper compounds made up almost one third of the total copper content at an application rate of 100 mg/kg, which was due to the predominant accumulation of organomineral copper compounds. At a rate of 100 mg/kg, the absolute content of these metal compounds increased compared to the clean soils by 54 times; at a rate of 300 mg/kg, this increase was 225 times (Table 3.11). Their share in the total metal content was doubled. Therefore, the contribution of organomineral copper compounds to the reserve of weakly bound metal compounds became comparable to the effect of compounds retained by carbonates and was predominant at the maximum level of soil contamination (Table 3.13).

Two years later, the content of mobile organomineral Cu compounds in the contaminated soil increased by 25–50% compared to the clean soil (Table 3.5). This increase was due to exchangeable copper ions. I. N. Antipov-Karataev (1947) was the first who noticed the exchange reactions with the participation of heavy metals (Hg, Pb, Cu) in soils; as early as 1940s, he showed that metal ions reacting with the soil participated in both exchangeable and nonexchangeble (specific) sorption. The capacity of metal ions for displacing calcium during ion exchange increased when their concentration in the solution interacting with chernozem samples increased, i.e., the metal-to-calcium ion ratio typical for chernozems was changed.

A possible reason for the increased involvement of soil organic matter in the reaction of soil with the metal could be the partial mobilization of humus under the effect of contamination, which was manifested in an increase in the share of fulvic acids in the humus composition and in the relative content of mobile organomineral copper compounds. This could be related to the decrease in the amount of exchange centers in the SEC of contaminated soils (Minkina et al., 2006).

The contribution of copper compounds weakly bound by Fe, Al, and Mn hydroxides to the pool of mobile copper compounds in the contaminated chernozem was slightly smaller than that of organomineral copper compounds. Their absolute content in the contaminated soil increased, compared to the clean soils, by almost two orders of magnitude, and their relative content increased by at least an order of magnitude (Table 3.11).

In the chestnut soil, the participation of exchangeable forms in the formation of the pool of weakly bound copper compounds in all experimental treatments was more important than in the chernozem. The share of exchangeable copper compounds was higher than that of copper complexes (Table 3.9). The low binding strength of copper with the solid phase of chestnut soil could be related to the lower content of humus, its higher fulvate content, the leaching of carbonates from soils, and a slightly higher alkalinity compared to chernozem (Table 3.1).

The conservation of general tendencies in the fractional and group composition of metal compounds in chernozem and chestnut soil was due to the similarity of their acid–base and ion-exchange properties, humus statuses, and particle-size distributions.

No stabilization in the proportions of different metal forms was observed during two years, which indicated low transformation rates of the applied metal compounds in soils (Minkina et al., 2008c).

When copper was added to soils together with other elements at low concentrations corresponding to MCLs for exchangeable metal forms, its mobility in the contaminated soils did not change (Table 3.10). In this case, the amount of free reactive centers in the sorption-exchange complex was larger compared to the amount of Cu^{2+}, Pb^{2+}, and Zn^{2+} ions present in the solution, and they had no appreciable mutual effect.

At higher application rates of metals, the mobility coefficient of copper increased by 1.3–2 times due to the proportional increase in the content of all mobile element fractions. The competition among the Cu^{2+}, Pb^{2+}, and Zn^{2+}

ions for reactive centers of the soil exchange complex was enhanced, as well as the displacement of weakly retained metal ions by more actively sorbed ions.

At the intensive increase in the content of complex copper compounds and those specifically sorbed by (hydr)oxides in the soils, the contribution of exchangeable and carbonate-bound fractions decreased (Table 3.13).

The ratio of weakly bound copper compounds did not change to the second year after contamination.

Group of strongly bound copper compounds. Nonsilicate minerals and especially organic substances strongly retained copper ions in the contaminated soils. These copper compounds became predominant in the strongly bound metal compounds (Tables 3.11. 3.13). The increase in their content in the contaminated soils was most important compared to other metals. The absolute content of copper strongly retained by silicates decreased by almost three times, although the absolute content of this fraction increased under soil contamination.

Lead Compounds

Group of Weakly Bound Lead Compounds

As in the case of copper, the contamination of soils with lead resulted in an increase in the absolute and relative contents of the metal compared to the background soils. This was mainly due to the formation of organomineral lead complexes. Their absolute content increased by 30 and 150 times at Pb application rates of 100 and 300 mg/kg, respectively. Therefore, the complexes of lead made up half the total content of its weakly bound compounds (Table 3.13).

The formation of lead complexes is responsive to both the increase in the application rate of the metal and its residence time in the soil. This was confirmed by the increase in the content of these compounds in chernozem in both cases (Table 3.5). For lead ions, this tendency was more manifested than for copper ions. This was due to the higher complexing capacity of humic acids with respect to lead ions and the higher stability of the forming complexes compared to organomineral copper complexes (Chemistry of Heavy Metals..., 1985).

It is suggested that the metals arrived in the soil because of contamination enhanced the complexation of organic substances, which could be accompanied by their partial destruction (Minkina et al., 2000a). These processes resulted in a decrease in the relative content of exchangeable metal forms during two years after contamination (Table 3.8). The mobility of lead in the chestnut soil was higher than that in the chernozem, which was largely due to the exchangeable metals forms (Table 3.9).

The increase in the portion of Pb complexes in the group composition of its compounds was accompanied by a decrease in the share of metal compounds specifically sorbed by carbonates. The content of specifically sorbed lead forms in soils decreased over the two-year period (Table 3.5).

Under conditions of polymetallic contamination, the mobility of lead in the soil also increased (Table 3.10). A three-fold increase in lead mobility was observed in the treatment where the content of added copper exceeded the lead application rate by a factor of 1.7 (Cu 55 + Pb 32 + Zn 100). Copper and lead ions actively participated in complexation; therefore, the simultaneous presence of copper more strongly affected the mobility of lead in the soil compared to zinc.

Group of strongly bound lead compounds. Metals retained by organic substances composed no less than half of the strongly bound lead compounds in the contaminated soils. This was due to the high capacity of metal ions to form chelates. Lead ions were active in the formation of bidentate complexes with the participation of functional groups of aromatic rings (Manceau, Marcus, Tamura, 2002). In the contaminated soils, the effect of nonsilicate Fe and Mn compounds on the strong fixation of lead ions was enhanced. The absolute content of these lead compounds in the contaminated soils increased by 15–42 times compared to their clean analogues, which resulted in a fourfold increase in their relative content in strongly bound compounds (Table 3.11).

Zinc Compounds

Group of Weakly Bound Zinc Compounds

The tendency toward an increase in metal mobility with increasing its application rate was more manifested for zinc than for copper and lead. The share of weakly bound zinc compounds in the contaminated chernozem increased by up 4 times compared to that in the clean chernozem and made up 40% of the total metal content.

The contamination of soils with zinc affected the qualitative composition of its more mobile compounds. The predominance of zinc compounds specifically sorbed by carbonates decreased. This fraction made up 88% in the group composition of zinc compounds in the clean soils. In the contaminated soils, the share of these compounds decreased by almost half and became comparable to the content of zinc compounds weakly bound by nonsilicate minerals. The absolute content of this metal fraction in the soils increased by 56–133 times (Table 3.11).

The content of zinc complexes in the contaminated soils also increased, although to a lesser extent than under contamination with copper and lead. This was related to the fact that zinc formed predominantly outer-sphere or unstable inner-sphere complexes with soil organic substances. Their effect on the mobility of metal in the contaminated soils decreased with time.

Another peculiar feature was also noted: the relative content of exchangeable metal forms in soils increased under contamination with zinc, in distinction from the soils contaminated with copper and lead. This fact agreed with the results of experiments on the sorption of heavy metals by ordinary chernozem, which showed that Zn^{2+} ions were mainly sorbed through ion exchange, and Cu^{2+} and Pb^{2+} ions were mainly bound by specific sorption (Minkina et al., 2005). Two years after the contamination with Zn^{2+} ions, the share of specifically sorbed zinc compounds increased. This could be partly due to the compounds previously occurred in the forms of exchangeable and complex compounds. The increase in the content of specifically sorbed compound by 20–31% resulted in an increase in zinc mobility in the contaminated soil.

In the contaminated chestnut soil, exchange processes even more significantly contributed to the retention of zinc than in chernozem (Table 3.9).

Under polymetallic contamination, the mobility of zinc in the soils studied exceeded the mobility of other metals (Table 3.10). This was indicative of a weak retention of Zn^{2+} ions competing with other metal ions (Minkina et al., 2005).

Group of strongly bound zinc compounds. Analogously to weakly bound zinc, its strongly bound compounds were most actively retained by nonsilicate iron and manganese compounds. The absolute content of zinc compounds bound to them increased by 17 times. At an application rate of 300 mg/kg, these compounds became predominant and exceeded the content of metal compounds in silicates. N.G. Zyrin et al. (1979) emphasized that zinc ions in neutral and slightly alkaline soils not only formed their own hydroxide precipitates but were also retained in large amounts by iron oxides and hydroxides. In comparison with humic acids, iron hydroxides sorbed 1.5 times more zinc ions and retained two times stronger (Brummer at al., 1983).

The content of zinc compounds strongly bound to organic matter was insignificant and varied only slightly under contamination, in distinction from the soils contaminated with copper and lead.

The determination of metal compounds in the soils studied under contamination in pot experiments showed that the metals could be arranged in the following series according to their retention by organic substances: Pb > Cu >> Zn. This series corresponds to the stabilities of their complexes with humic substances (Andersson, 1977). This series also coincides with changes in their ionic radii (Pb > Cu > Zn), which suggests that the size of metal ions affects their retention by organic substances (Minkina et al., 2000b).

The participation of major soil components, organic substances, and nonsilicate iron and manganese compounds in the retention of metals under soil contamination is shown in Table 3.14.

In the initial state, organic substances and nonsilicate iron and manganese compounds were active in the strong retention of metals; the participation of organic substances was more significant for lead and copper ions, and iron and manganese oxides and hydroxides were more involved in the retention of zinc ions.

This tendency remained in the soils contaminated with metals: copper and lead ions introduced in the soil were predominantly bound by organic substances, and zinc ions by nonsilicate iron and manganese compounds. Both soil components participated in the strong and weak retention of metals; the ratio of these compounds in the contaminated soils changed toward an increase in the share of weakly bound compounds. It was noted that the low capacity of organic substances to strongly fix zinc ions further decreased in the contaminated soils.

Thus, the predominance of strongly bound metal compounds (88–95% of the total amount) in the initial soils was mainly ensured by their fixation in the lattices of silicate minerals (56–83% of the total amount). The mobility of Cu, Pb, and Zn in the soils was low (5–12%) and mainly due to metal compounds retained by carbonates.

The contamination of ordinary chernozem from the Rostov oblast disturbed its natural metal ratio. The addition of metals to the soil (at an application rate of 100–300 mg/kg) increased the content of all Cu, Pb, and Zn compounds, but their proportions abruptly changed toward an increase in the content of mobile compounds (to 30–40% of the total amount). This increase was due to exchangeable, complex, and specifically sorbed compounds, which were mutually related by sequential transformations. A dynamic formation–transformation equilibrium was established between different forms of mobile metals in the soils. Predominant mobile compounds were specifically sorbed ones, which could be considered as transitional to strongly bound compounds. In the second years after the application of zinc, their content increased, probably due to the compounds previously occurred in the forms of exchangeable and complex compounds.

Complexes were formed more rapidly than other compounds for all metals. The content of copper and lead complexes increased with time.

The ion-exchange sorption was more typical for zinc than for copper and lead.

Organic substances and nonsilicate Fe, Al, and Mn minerals were the main retaining agents for both strongly and weakly bound compounds. The fixation of applied metals in the lattices of silicate minerals was insignificant.

A larger role of exchange processes in the transformation of heavy metal compounds was observed in chestnut soil compared to chernozem.

Among the metals considered, zinc was the least strongly retained and the weakest competitor for adsorption sites; its mobility increased in the presence of copper and lead.

Soil properties affected the mobility of heavy metals, but soils themselves underwent changes because of contamination. Their capacity to strongly fix metal decreased with increasing rate of heavy metals. The composition of metal compounds in soils can serve as a criterion of the ecological (barrier) function of soils.

3.2.2. Transformation of Zinc Compounds under Field Contamination Conditions

A field experiment was established to assess the effect of the spatial and temporal variation in soil-forming factors on the status of heavy metals in the soil. The rates and forms of application of zinc and lead in the field experiment were analogous to those used in the treatments of the pot experiment with the highest degree of contamination by these metals.

3.2.2.1. Procedure of Field Experiment

A long-term (1999–2004) field experiment was established on a thick calcareous low-humus clay loamy ordinary chernozem developed on loess-like loams. The properties of this soil are given in Table 3.15.

Table 3.15. Chemical and physical properties of the studied soil (0 to 20-cm layer)

Soil	Humus, %	CaCO$_3$, %	pH	N-NO$_3$, mg/100 g	P$_2$O$_5$ mob., mg/100 g	K$_2$O exch., mg/100 g	Exchangeable ions, meq/100 g Ca^{2+}	Exchangeable ions, meq/100 g Mg^{2+}	<0.01 mm, %	<0.001 mm, %
Calcareous ordinary chernozem, (Aksaiskii region)	3,8	0,15	7,5	0,9	6,0	35,6	30	4,5	58,0	34,5

The soil was contaminated with zinc and lead salts in 1999. The metals were applied separately as dry acetate salts to the plow (0- to 20-cm) horizon in fall. The application rates were 300 mg/kg for Zn and 100 mg/kg for Pb, which corresponded to 3 MLC for these metals in the soil.

Spring barley (*Hordeum sativum distichum*), cultivar Odesskii-100, was planted in experimental plots in 2000. The crop management practice recommended for this zone was used. Samples of soil (from the 0- to 2-cm layer) and plants were taken at the complete maturity stage of barley (Fig. 3.1) during three years after the beginning of the experiment for studying the effect and aftereffect of heavy metals.

The establishment of experiments, observations, recordings, and sampling of soils and plants were performed in accordance with procedures of field experiments (Dospekhov, 1968). Experiments were conducted in triplicate.

Figure 3.1. Field experiment at the complete maturity stage of barley.

3.2.2.2. Results of Field Experiment

The content of metal in the control plots (the treatment without application of metals) during three years of studies was 65–68 mg/kg for Zn and 24–28 mg/kg for Pb (Table 3.16), which reflected the variation of this parameter for soils of the pot experiment. The content of metals in three fractions of weakly bound zinc and lead compounds characterized changes in these parameters with time (Table 3.17).

Table 3.16. The total content and Zn and Pb weakly bound compounds in the chernozem during 3 year after contamination, mg/kg (n=9)

Experimental treatments	Exchangeable compounds			Complex compounds			Specifically sorbed compounds			Total content		
	1 year	2 year	3 year	1 year	2 year	3 year	1 year	2 year	3 year	1 year	2 year	3 year
Zn												
Without metal addition	0,6	0,6	0,6	0,4	0,5	0,4	6,5	6,8	6,9	68	65	67
Metal	33,0	27,6	26,1	27,9	23,7	24,4	52,3	69,8	76,6	356	349	352
LSD$_{0.95}$	6,4	8,0	2,3	1,4	2,01	2,2	10,3	9,9	5,1	16	16	17
Pb												
Without metal addition	0,8	0,9	1,0	0,3	0,3	0,1	2,4	2,5	2,2	24	24	28
Metal	12,8	10,8	8,7	6,0	12,8	14,7	22,9	21,7	18,3	110	101	100
LSD$_{0.95}$	1,4	4,0	1,1	1,3	2,3	1,6	9,9	10,5	4,0	6	8	8

The total metal content in the ordinary chernozem contaminated under field experimental conditions increased by 5.2 times for zinc and by 4.6 times for lead. As was noted above, in clean soils, metals predominantly occurred in the strongly bound state (85–88% of the total amount). The hazard of soil contamination with metals is related not only to the increase in metal content, but also to the increase in metal mobility. Under contamination, the mobility coefficient Km of metals increased by 5 times for Zn and by 3 times for Pb (Table 3.17).

In the clean soils, the share of weakly bound metal compounds was no more than 12–15%; in the contaminated soils, they made up almost one third and one half of the total metal content for zinc and lead, respectively, predominantly due to exchangeable and complex metal compounds. The value of Km for lead was higher than for zinc. This was due to the high content of exchangeable Pb compounds. Although the distributions of Zn and Pb among different forms in the clean and contaminated soils were similar (specifically sorbed > exchangeable > complex), the contents of metals in these three fractions became more similar under contamination. The artificial contamination of the soils resulted in an increase in the content of exchangeable Zn and Pb above their MCL levels, which remained for several years (Table 3.16).

The share of weakly bound lead and zinc compounds remained almost constant with time, but the ratio of their fractions was disturbed. Already in the second and third years after the contamination of chernozem, the accumulation of specifically sorbed zinc occurred due to exchangeable and complex compounds, as well as the accumulation of complex lead compounds due to exchangeable forms. To this time, a decrease in the accumulation of specifically sorbed Pb forms in weakly bound compounds became apparent, which could be related to their transition to a more strongly bound state.

Thus, the contamination of ordinary chernozem with zinc and lead under field experimental conditions was accompanied by an increase in metal mobility: the relative content of weakly bound compounds increased by 2–3 times. Changes in the contents of different fractions occurred within this group: the share of exchangeable and complex zinc and lead compounds increased and the share of their specifically sorbed forms decreased.

No equilibrium was reached in the system of metal compounds in chernozem during three years after the application of zinc and lead. The initial transformation stage of the added metal ions was mainly related to the partial transition of exchangeable element forms to complex forms for lead and to the transition from exchangeable forms to specifically sorbed forms for zinc.

Table 3.17. Group composition and mobility (Km) parameters of Pb and Zn compounds in polluted chernozem

Experimental treatments	WB/SB *			Exchangeable/ complex/ specifically sorbed **			K_m		
	1 year	2 year	3 year	1 year	2 year	3 year	1 year	2 year	3 year
Zn									
Without metal addition	12/88	12/88	12/88	8/5/87	8/6/86	8/5/87	0,1	0,1	0,1
Metal	32/68	35/65	36/64	29/25/46	23/19/58	21/19/60	0,5	0,5	0,6
Pb									
Without metal addition	15/85	15/85	12/88	23/9/68	24/8/68	30/3/67	0,2	0,2	0,1
Metal	38/62	45/55	42/58	31/14/55	24/28/48	21/35/44	0,6	0,8	0,7

* The WB/SB values are given in percent of the total content; ** the contents of the exchangeable/complex/specifically sorbed metals are given in percent of the content of weakly bound metals.

3.2.3. Transformation of Copper, Zinc, and Lead Compounds in the Soils of Monitoring Plots under Aerotechnogenic Contamination

Rostov oblast is included in the list of regions with unfavorable ecological situation (Zakrutkin, Ryshkov, 1997). About 40% of population inhabits under contaminated atmospheric conditions. The Novocherkassk power station is the most powerful source of atmospheric contamination.

The emission of this enterprise makes up 1% of the total emission in the Russian Federation and 58% of its volume in Rostov oblast, where the contribution of Novocherkassk is 99% (Kizel'shtein et al., 1990; State Report..., 1998; Ecology of Novocherkassk, 2001; Motuzova, Bezuglova, 2007). According to the amount of pollutant aerosols, Novocherkassk was repeatedly included in the priority list of Russian cities with the highest pollution level. The Novocherkassk power station is located only at 7.5 km to the southeast of the city. The sanitary zone is not observed; the wind diagram is not taken into consideration, and 99% of emissions from the power station fall onto the residential areas.

The accumulation of heavy metals in soils depends on their input and removal. The imbalance of heavy metal fluxes results in their accumulation in soils. Its consequences depend on the technogenic load and soil buffering capacity with respect to heavy metals. Long-term stationary observations of technogenically disturbed soils performed in the zone affected by the Novocherkassk power station are a valuable source of information.

3.2.3.1. Procedure of Monitoring Observations

Monitoring plots were established in 2000. They were located at different distances (1 to 20 km) from the Novocherkassk power station (Fig. 3.2) and confined to the sites of air sampling performed at the project development stage of the organization and arrangement of the sanitary zone for the northern industrial region of Novocherkassk (Fig. 3.2, plots 1–3, 5–7) (Anthropogenic Effect of Emissions..., 1995; Report, 1995; Ecological Certificate of the City of Novocherkassk, 1995; Toporskaya, Danilova, 1997).

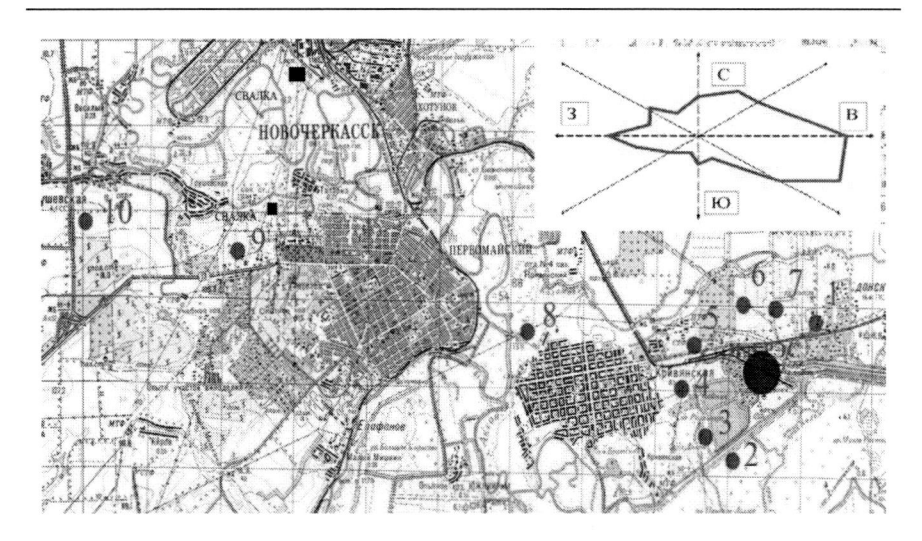

Composition of Copper, Zinc, and Lead Compounds... 85

Figure 3.2. Sketch map of monitoring plots in the zone affected by the Novocherkassk power station.

Direction and distance from the Novocherkassk power station

Plot 1	1 km to the northeast;
Plot 2	3 km to the southwest;
Plot 3	*2.7 km to the southwest;*
Plot 4	1.6 km to the northwest;
Plot 5	1.2 km to the northwest;
Plot 6	2.0 km to the north-northwest;
Plot 7	1.5 km to the north;
Plot 8	5 km to the northwest;
Plot 9	15 km to the northwest.
Plot 10	20 km to the northwest.

In accordance with the wind diagram, the so-called general direction was determined: a straight line from the contamination source through the residential zones of the city of Novocherkassk and the village of Krivyanskaya. Along the general direction, soil samples were taken in monitoring plots 4, 8, 9, and 10.

Monitoring plots should also be established on virgin or fallow areas, where the soil was not tilled and its layers were not mixed (Fig. 3.3).

Figure 3.3. Monitoring plots in the zone affected by the Novocherkassk power station: (A) plot 4; (B) plot 8.

Samples were taken from a depth of 0–20 cm. The properties of monitoring plot soils are given in Table 3.18.

Most soils of monitoring plots were chernozems, and the presumably uncontaminated soils of plots 9 and 10 served as control soils for their assessment. The calcareous low-humus sandy alluvial meadow soil (plot 2), which had a coarse texture and a low CEC, and the low-humus sandy clay floodplain meadow-chernozemic soil (plot 3) with a high CEC differed from the control soils. The differences between these soils and the control soils were taken into account in the discussion of results.

3.2.3.2. Results of Monitoring Observations

The contents of exchangeable Zn and Cu in plot 10 and that of exchangeable Pb in plot 9 were the lowest: about 1.0 mg/kg (Table 3.19).

The total content of metals in the clean soils of the areas controlled (plots 9, 10) was mainly (80–89%) ensured by strongly bound metal compounds (Table 3.20). The mobility coefficients (Km) of metals varied in the range 0.1–0.3. The group of weakly bound heavy-metal compounds mainly consisted of specifically sorbed forms (72–76% of weakly bound compounds). The metal compounds in this group formed the following series: specifically sorbed > exchangeable ≥ complex.

An increase in the total content of Cu, Pb, and Zn was observed in the soils of all monitoring plots, except the most remote ones (at 15–20 km), compared to the background levels of metals in the soils of Novocherkassk (Ecology of Novocherkassk, 2001).

The highest contents of metals exceeding the MCLs for total Cu, Zn, and Pb were observed in the surface soil layers of the closest monitoring plots to the contamination source with account for the wind diagram (plots 4, 5, 8) and that of plot 6. In the soil of plot 1, the total contents of lead and zinc exceeded their MCLs (Table 3.19).

At the same time, the soils of plot 1 and especially plots 2 and 3 were less contaminated, although they were located closely to the contamination source. This was related to the fact that these plots were beyond the zone of predominant wind directions.

Table 3.18. Chemical and physical properties of the soils around the Novocherkassk Power Station (0 to 20-cm layer)

№ of the plot	Soil	<0.01 mm, %	<0.001 mm, %	Humus, %	pH	CaCO₃, %	NH_4^+ mg/100 g	P_2O_5	K_2O	$Ca^{2+}+Mg^{2+}$ meq/100 g	CEC
1	Medium-thick medium-humus clay loamy calcareous ordinary chernozem on loess-like loams	52,3	29,6	4,2	7,6	0,5	2,8	3,6	39,9	32,3	35,2
2	Low-humus sandy calcareous alluvial meadow soil on alluvial deposits	5,9	2,9	3,1	7,5	0,3	2,4	1,5	20,9	10,0	10,3
3	Low-humus sandy clay floodplain meadow-chernozemic soil on alluvial deposits	63,4	36,8	4,6	7,2	0,3	2,1	4,5	34,7	40,2	44,3
4	Medium-thick medium-humus clay loamy calcareous ordinary chernozem on loess-like loams	55,3	30,9	4,5	7,4	0,8	2,9	4,0	30,4	32,1	33,2
5	Medium-thick medium-humus clay loamy calcareous ordinary chernozem on loess-like loams	56,3	30,8	4,3	7,4	0,7	2,4	3,0	37,3	35,8	37,6

6	Medium-thick medium-humus clay loamy meadow-chernozemic soil on loess-like loams	58,8	34,9	4,1	7,6	0,8	3,6	3,2	35,1	30,3	32,0
7	Medium-thick medium-humus clay loamy calcareous ordinary chernozem on loess-like loams	53,7	30,3	4,2	7,5	0,6	3,0	2,6	48,5	31,1	33,2
8	Medium-thick medium-humus clay loamy meadow-chernozemic soil on loess-like loams	60,0	32,4	4,8	7,2	0,3	2,3	4,4	33,0	45,6	49,9
9	Medium-thick medium-humus clay loamy calcareous ordinary chernozem on loess-like loams	54,3	31,8	4,3	7,6	0,7	2,2	3,7	32,8	33,6	34,4
10	Medium-thick medium-humus clay loamy calcareous ordinary chernozem on loess-like loams	55,1	30,0	4,5	7,7	0,7	3,9	3,8	40,7	35,0	37,1

Table 3.19. The total content and Cu, Zn and Pb weakly bound compounds in the soils around the Novocherkassk Power Station, mg/kg

№ of the plot, distances (km) and directions from the Power station	Weakly bound compounds			Specifically sorbed compounds	Sum of the weakly bound compounds	Total content
	Exchangeable compounds	Complex compounds				
Cu						
1,0 NE	2,4	2,2		9,7	14,3	51
3,0 SW	3,6	1,3		6,1	11,0	44
2,7 SW	1,8	1,5		5,8	9,1	50
1,6 NW	4,5	4,8		14,9	24,2	75
1,2 NW	3,4	5,4		12,6	21,4	60
2,0 NNW	3,6	4,5		12,2	20,3	58
1,5 N	1,3	1,9		7,1	10,3	39
5,0 NW	3,2	4,3		11,9	19,4	59
15,0 NW	1,0	0,7		5,5	7,2	37*
20,0 NW	0,7	0,7		3,5	4,9	37*
LSD$_{0,95}$	1,1	0,9		1,0	2,2	5
MCLs	3,0					55
Pb						
1,0 NE	3,5	2,9		8,2	14,6	40
3,0 SW	2,0	0,2		2,2	4,4	20
2,7 SW	1,7	1,4		3,8	6,8	28
1,6 NW	6,6	3,3		11,5	21,4	65
1,2 NW	6,1	4,4		14,0	24,5	60

Table 3.19. (Continued).

№ of the plot, distances (km) and directions from the Power station	Weakly bound compounds			Specifically sorbed compounds	Sum of the weakly bound compounds	Total content
	Exchangeable compounds	Complex compounds			
2,0 NNW	4,6	4,3	12,3	21,2	61
1,5 N	3,3	1,8	7,3	12,4	32
5,0 NW	2,9	2,8	5,3	11,0	43
15,0 NW	1,0	0,4	3,6	5,0	25*
20,0 NW	3,0	2,4	6,4	11,8	37
LSD$_{0,95}$	1,1	0,7	1,0	0,9	3
MCLs	6,0				32
Zn					
1,0 NE	10,6	3,9	28,6	43,1	106
3,0 SW	11,4	1,0	17,9	30,3	79
2,7 SW	4,4	4,4	19,7	28,5	99
1,6 NW	15,5	3,6	23,8	42,9	107
1,2 NW	24,8	3,1	31,2	59,1	140
2,0 NNW	12,7	1,7	30,1	44,5	115
1,5 N	5,2	1,2	17,8	24,2	92
5,0 NW	14,2	3,3	21,8	39,3	113
15,0 NW	1,9	1,1	7,6	10,6	80*
20,0 NW	1,1	0,8	6,1	8,0	72
LSD$_{0,95}$	0,8	1,0	1,1	3,7	9,3
MCLs	23,0				100,0

* Conformity to the background level of metal in the soil.

Table 3.20. Group composition and mobility (Km) parameters of Cu, Pb and Zn compounds in the soils around the Novocherkassk Power Station

№ of the plot, distances (km) and directions from the Power station	WB/SB *	Exchangeable/ complex/ specifically sorbed **	K_m
Cu			
1,0 NE	28/72	17/15/68	0,4
3,0 SW	25/75	33/12/55	0,3
2,7 SW	18/82	20/16/64	0,2
1,6 NW	32/68	19/20/61	0,5
1,2 NW	36/64	16/25/59	0,6
2,0 NNW	35/65	18/22/60	0,5
1,5 N	26/74	13/18/69	0,4
5,0 NW	33/67	16/22/62	0,5
15,0 NW	19/81	14/10/76	0,2
20,0 NW	13/87	14/14/72	0,2
Pb			
1,0 NE	37/63	24/20/56	0,6
3,0 SW	22/78	45/5/50	0,3
2,7 SW	24/76	25/20/55	0,3
1,6 NW	33/67	31/15/54	0,5
1,2 NW	41/59	25/18/57	0,7
2,0 NNW	35/65	22/20/58	0,5
1,5 N	39/61	27/14/59	0,6
5,0 NW	26/74	26/25/49	0,4
15,0 NW	20/80	20/8/72	0,3
20,0 NW	32/68	25/20/55	0,5
Zn			
1,0 NE	41/59	24/9/67	0,7
3,0 SW	38/62	38/3/59	0,6

Table 3.20. (Continued).

№ of the plot, distances (km) and directions from the Power station	WB/SB *	Exchangeable/ complex/ specifically sorbed **	K_m
2,7 SW	29/71	16/15/69	0,4
1,6 NW	40/60	36/8/56	0,7
1,2 NW	42/58	42/5/53	0,7
2,0 NNW	39/61	29/4/67	0,6
1,5 N	26/74	21/5/74	0,4
5,0 NW	35/65	36/8/56	0,5
15,0 NW	13/87	18/10/72	0,2
20,0 NW	11/89	14/10/76	0,1

* The WB/SB values are given in percent of the total content; ** the contents of the exchangeable/complex/specifically sorbed metals are given in percent of the content of weakly bound metals.

The major part of pollutants settles within 5 km from the contamination source along the general direction. The content of heavy metals gradually decreases with increasing distance and approaches the background values in the most remote areas (plots 9, 10). The soil of plot 10, which is located at 350 m from a motor road, is characterized by an increased content of lead, which could be due to the effect of vehicle exhaust gases.

L. K. Kazakov (1977) studied the soils affected by emissions from several big power stations in the central Russia and found that the maximum ecological disturbances in landscapes are observed within a distance of 4 km from te emission source. A zone of local centers of significant disturbances is located within 4–8 km from the power stations, and a zone of local moderate and low disturbances is at 8 to 15 km.

In the contaminated soils, the share of weakly bound metal compounds increased appreciably. A significant increase in the content of mobile metal compounds occurred in the soils of plots 4, 5, 6, and 8 for copper; plots 5 and 6 for lead; and plot 5 for zinc (Table 3.19). In the soils of the plots close to the Novocherkassk power station, the shares of mobile Cu, Pb, and Zn compounds increased by 2.8, 2.0, and 3.8 times, respectively (Table 3.20). The changes in proportions of heavy-metal groups entailed a significant reorganization of their fractional composition. The increase in the share of more mobile compounds with increasing long-term accumulation of metal was the main tendency of the changes observed.

The absolute content of all groups and fractions of metal compounds increased in the contaminated soils; the absolute content of metals in the residual fraction changed to a lesser extent (Table 3.21). The relative content of metals in the residual fraction decreased by 20–30% (Tables 3.22, 3.23), which ensured a decrease in the total content of strongly bound metal compounds. These data agreed with the previous results (Section 3.2.1.2).

The mobility of metals in soils depends on the buffering properties of soil with respect to these elements. The plot ratings and degrees of buffering of soils with respect to metals calculated by the Il'in procedure (1995, 2001) with account for the effects of different factors on the buffering capacity of soils in plots are given in Table 3.24.

Table 3.21. Group composition of Cu, Pb and Zn in the soils around the Novocherkassk Power Station, mg/kg

№ of the plot	Weakly bound compounds (WB) Exchangeable AAB/MgCl₂	Complex	Specifically sorbed on carbonates	on Fe and Mn (hydr)oxides	Strongly bound compounds (SB) With organic matter	With Fe and Mn (hydr)oxides	With silicates	Sum of fractions
Cu								
1	2,4/0,7	2,2	6,7	3,0	9,0	6,7	26,8	55,1
2	3,6/0,9	1,3	3,1	3,0	2,0	3,3	20,9	34,5
3	1,8/0,9	1,5	2,0	3,8	12,4	8,1	32,0	60,7
4	4,5/2,3	4,8	6,0	8,9	16,4	14,5	35,5	88,4
5	3,4/2,0	5,4	7,3	5,3	13,7	10,1	28,4	72,4
6	3,6/1,0	4,5	6,2	6,0	16,0	12,8	17,5	64,0
7	1,3/1,3	1,9	2,1	5,0	10,7	7,2	25,5	53,7
8	3,2/1,8	4,3	2,8	9,1	9,5	7,0	34,5	69,0
9	1,0/1,0	0,7	3,5	2,0	6,3	5,0	22,0	40,5
10	0,7/0,4	0,7	2,5	1,0	3,3	3,0	28,7	39,6
Pb								
1	3,5/0,9	2,9	6,3	1,9	5,1	4,0	25,0	46,1
2	2,0/0,9	0,2	1,7	0,5	1,5	5,3	6,6	16,7
3	1,7/0,3	1,4	2,3	1,5	10,5	6,4	10,9	33,3
4	6,6/3,2	3,3	8,5	3,0	13,5	7,4	25,5	64,4
5	6,1/2,8	4,4	9,3	4,7	11,7	7,4	24,8	65,1
6	4,6/1,5	4,3	7,5	4,8	18,0	10,8	17,5	64,4
7	3,3/0,5	1,8	5,0	2,3	11,0	6,2	11,5	38,3
8	2,9/1,0	2,8	2,5	2,8	12,1	5,4	12,1	38,7
9	1,0/0,3	0,4	1,5	2,1	5,0	1,8	10,3	21,4

№ of the plot	Weakly bound compounds (WB)			Specifically sorbed		Strongly bound compounds (SB)			Sum of fractions
	Exchangeable AAB/MgCl$_2$	Complex	on carbonates	on Fe and Mn (hydr)oxides	With organic matter	With Fe and Mn (hydr)oxides	With silicates	
10	3,0/0,7	2,4	3,9	2,5	6,0	4,1	16,4	36,0
Zn								
1	10,5/4,1	3,9	16,0	12,6	5,8	9,6	60,7	112,7
2	11,4/5,5	1,0	11,1	6,8	2,5	16,0	45,9	88,8
3	4,4/2,0	4,4	10,0	9,7	8,4	14,7	56,5	105,7
4	15,5/8,2	3,6	14,8	9,0	4,4	10,4	50,0	100,4
5	24,8/10,8	3,1	17,3	13,9	6,8	19,1	70,3	141,3
6	12,7/6,3	1,7	14,8	15,3	6,5	20,8	56,8	122,2
7	5,2/3,3	1,2	11,7	6,1	7,5	7,2	41,6	78,6
8	14,2/4,0	3,3	12,8	9,0	6,0	13,1	51,8	100,0
9	1,9/1,3	1,1	6,0	1,6	5,2	4,0	54,4	73,6
10	1,1/0,5	0,8	4,0	2,1	4,1	6,4	61,9	79,8

Table 3.22. Relative contents of Cu, Pb and Zn compounds in the soils around the Novocherkassk Power Station, % of the sum of the fractions

№ of the plot	Weakly bound compounds (WB)			Strongly bound compounds (SB)			WB/SB
	Exchangeable, MgCl₂	Complex	Specifically sorbed		With organic matter	With Fe and Mn (hydr)oxides	With silicates	
			On carbonates	on Fe and Mn (hydr)oxides				

Cu

№	Exch. MgCl₂	Complex	On carbonates	on Fe/Mn	Org. matter	Fe/Mn	Silicates	WB/SB
1	1,3	4,0	12,2	5,4	16,3	12,2	48,6	23/77
2	2,6	3,8	8,9	8,7	5,8	9,6	60,6	24/76
3	1,5	2,5	3,3	6,3	20,4	13,3	52,7	14/86
4	2,6	5,4	6,8	10,1	18,6	16,4	40,2	25/75
5	2,8	7,5	10,1	7,3	18,9	14,0	39,2	28/72
6	1,6	7,0	9,7	9,4	25,0	20,0	27,3	28/72
7	2,4	3,5	3,9	9,3	19,9	13,4	47,5	19/81
8	2,6	6,2	4,1	13,2	13,8	10,1	50,0	26/74
9	2,5	1,7	8,6	4,9	15,6	12,3	54,3	18/82
10	0,9	1,6	5,9	2,3	7,7	7,0	67,4	11/89

Pb

№	Exch. MgCl₂	Complex	On carbonates	on Fe/Mn	Org. matter	Fe/Mn	Silicates	WB/SB
1	2,0	6,3	13,7	4,1	11,1	8,7	54,2	26/74
2	5,4	1,2	10,1	3,0	8,9	31,5	39,3	20/80
3	0,9	4,2	6,9	4,5	31,5	19,2	32,7	17/83
4	5,0	5,1	13,2	4,7	21,0	11,5	39,6	28/72
5	4,3	6,8	14,3	7,2	18,0	11,4	38,1	33/67
6	2,3	6,7	11,6	7,5	28,0	16,8	27,2	28/72
7	1,3	4,7	13,1	6,0	28,7	16,2	30,0	25/75

	Weakly bound compounds (WB)					Strongly bound compounds (SB)				
			Specifically sorbed						
№ of the plot	Exchangeable, MgCl$_2$	Complex	On carbonates	on Fe and Mn (hydr)oxides	With organic matter	With Fe and Mn (hydr)oxides	With silicates	WB/SB
8	2,6	7,2	6,5	7,2	31,3	14,0	31,3	24/76
9	1,4	1,9	7,0	9,8	23,4	8,4	48,1	20/80
10	1,9	6,7	10,8	6,9	16,7	11,4	45,6	26/74
Zn								
1	3,6	3,5	14,2	11,2	5,1	8,5	53,9	32/68
2	6,2	1,1	12,5	7,7	2,8	18,0	51,7	27/73
3	1,9	4,2	9,5	9,2	7,9	13,9	53,5	25/75
4	8,2	3,6	14,7	9,0	4,4	10,4	49,8	35/65
5	7,6	2,2	12,2	9,8	4,8	13,5	49,8	32/68
6	5,2	1,4	12,1	12,5	5,3	17,0	46,5	31/69
7	4,2	1,5	14,9	7,8	9,5	9,2	52,9	28/72
8	4,0	3,3	12,8	9,0	6,0	13,1	51,8	29/71
9	1,8	1,5	8,2	2,2	7,1	5,4	73,9	14/86
10	0,6	1,0	5,0	2,6	5,1	8,0	77,6	9/91

Table 3.23. Relative contents of Cu, Pb and Zn forms in groups of weakly and strongly bound compounds in the soils around the Novocherkassk Power Station

№ of the plot	Weakly bound compounds (WB)			Specifically sorbed		Strongly bound compounds (SB)		
	Exchangeable, MgCl$_2$	Complex	on carbonates	on Fe and Mn (hydr)oxides	With organic matter	With Fe and Mn (hydr)oxides	With silicates
Cu							
1	6	17	53	24	21	16	63
2	11	16	37	36	8	13	80
3	11	18	24	46	24	15	61
4	10	22	27	40	25	22	53
5	10	27	37	27	26	19	54
6	6	25	35	34	35	28	38
7	13	18	20	49	25	17	59
8	10	24	16	51	19	14	68
9	14	10	49	28	19	15	66
10	9	15	54	22	9	9	82
Pb							
1	8	24	53	16	15	12	73
2	27	6	52	19	11	40	49
3	5	25	42	27	38	23	39
4	18	18	47	17	29	16	55
5	13	21	44	22	27	17	56
6	8	24	41	27	39	23	38
7	5	19	52	24	38	22	40
8	11	31	27	31	41	18	41
9	7	9	35	49	29	11	60
10	7	25	41	26	23	15	62
Zn							

№ of the plot	Weakly bound compounds (WB)			Specifically sorbed		Strongly bound compounds (SB)		
	Exchangeable, MgCl$_2$	Complex	on carbonates	on Fe and Mn (hydr)oxides	With organic matter	With Fe and Mn (hydr)oxides	With silicates
1	11	11	44	34	8	13	80
2	23	4	45	28	4	25	71
3	8	17	38	37	11	18	71
4	23	10	42	25	7	16	77
5	24	7	38	31	7	20	73
6	17	4	39	40	8	25	68
7	15	5	52	27	13	13	74
8	14	11	44	31	8	18	73
9	13	11	60	16	8	6	86
10	7	11	54	28	6	9	85

Table 3.24. Buffering capacity of soils for heavy metals

№ of the plot	Soil	Points	Buffering capacity for heavy metals
1	Medium-thick medium-humus clay loamy calcareous ordinary chernozem on loess-like loams	38	increased
2	Low-humus sandy calcareous alluvial meadow soil on alluvial deposits	22	medium
3	Low-humus sandy clay floodplain meadow-chernozemic soil on alluvial deposits	42	high
4	Medium-thick medium-humus clay loamy calcareous ordinary chernozem on loess-like loams	37	increased
5	Medium-thick medium-humus clay loamy calcareous ordinary chernozem on loess-like loams	37	increased
6	Medium-thick medium-humus clay loamy meadow-chernozemic soil on loess-like loams	38	increased
7	Medium-thick medium-humus clay loamy calcareous ordinary chernozem on loess-like loams	37	increased
8	Medium-thick medium-humus clay loamy meadow-chernozemic soil on loess-like loams	42	high
9	Medium-thick medium-humus clay loamy calcareous ordinary chernozem on loess-like loams	40	increased
10	Medium-thick medium-humus clay loamy calcareous ordinary chernozem on loess-like loams	40	increased

The largest differences in buffering properties were revealed for the soils strongly differing in particle-size distribution. The sandy clay meadow-chernozemic soils (plot 3) had the highest buffering capacity; the sandy alluvial meadow soil (plot 2) had the lowest buffering capacity. The highest content of metals in the exchangeable state was found in the sandy alluvial soil compared to other soils, as well as an insignificant content of their complex compounds. This was related to the low sorption capacities of these soils characterized by the low contents of the clay fraction and organic matter (Table 3.18), whereupon the surface sorption of ions by soil components was the main mechanism of metal binding. The high mobility of metals in the sandy alluvial meadow soil could result in the migration of metals from these soils to ground and surface waters.

In the meadow-chernozemic soils (plots 3, 6, 8), organic substances actively interacted with heavy metals (especially Cu and Pb); the highest content of organomineral metal complexes was observed in these soils. This was especially manifested for copper and lead, which are active complexing agents.

In accordance with the increasing hazard of contamination at similar metal loads, the soils formed the following decreasing series: meadow-chernozemic soils > ordinary chernozems > alluvial meadow soils.

The high fertility level of chernozemic soils mitigates the negative effect of the long-term accumulation of pollutants and prevents the degradation of the soil and plant cover in the zone of technogenic impact. The soils of Novocherkassk suburbs can partially inactivate metals arrived together with aerosol emissions. However, this capacity of soil is limited. This was confirmed by a significant increase in the content of mobile metal compounds in soils under contamination conditions. In the soils of monitoring plots with the highest technogenic load, the share of weakly bound Cu, Zn, and Pb compounds increased to the greatest extent (Table 3.22). The high mobility of metals in these soils was due to the long-term accumulation of metals arrived with aerosol emissions from the Novocherkassk power station and the polymetal character of contamination.

The regular control of the soil status in the monitoring plots showed that the content of metal compounds in the soils insignificantly changed in the last 8 years (from 2000 to 2007). In 2007, the content of exchangeable Cu forms in the soils located at 5–20 km from the power station increased by 1.8 times compared to 1992; the content of exchangeable Zn and Pb compounds changed to a lesser extent. Nonetheless, the regular control of the soils subject to contamination is necessary. A tendency toward an increase in metal mobility in soils is clearly manifested, and the buffering capacity of soils for metals is limited.

The state of separate metals in soils of the technogenic zone near the Novocherkassk power station is considered below.

COPPER COMPOUNDS IN THE SOILS OF MONITORING PLOTS

Group of Weakly Bound Copper Compounds

In the clean chernozem, copper is strongly fixed by soil components, and only one eighth to one fifth of the metal amount is in the potentially mobile state (plots 9 and 10). The carbonate-bound fraction is predominant in the weakly bound metal compounds. The absolute and relative contents of exchangeable, complex copper forms, and those bound to Fe and Mn (hydr)oxides are also low (Tables 3.21, 3.23).

Under technogenic contamination, the share of weakly bound copper compounds can reach one third of the total copper content in the soil. The content of complex copper compounds and those bound to Fe and Mn (hydr)oxides increases to a greatest extent (to 8–9 times) with decreasing distance from the source of metal emissions. The content of exchangeable metal forms (extractable by 1 M $MgCl_2$) and their reserve (extracted by a 1 N ABB solution) increase by 2.5–6.3 times. The lowest increase is observed for the content of copper compounds bound to carbonates. Accordingly, the proportions of weakly bound metal compounds change: the increase in the share of complexes and compounds bound to Fe and Mn (hydr)oxides was accompanied by a decrease in carbonate-bound metal compounds, the content of exchangeable metal form remaining almost constant (Table 3.23). In some experimental treatments, the share of copper in Fe and Mn (hydr)oxides can be equal to or slightly higher than its content in carbonates.

Group of Strongly Bound Copper Compounds

In the clean soils, the following distribution of strongly bound metal compounds is observed: bound to silicates > bound to organic matter > bound to Fe and Mn (hydr)oxides. This series remains for the contaminated soils. An exception is provided by the alluvial meadow soil, where copper ions are more actively fixed by Fe and Mn oxides and hydroxides than by organic substances.

When the technogenic load increases, the absolute contents of the metal retained by organic substances and Fe and Mn (hydr)oxides increase similarly (by 5.5 times). The content of the element in the residual fraction changes to a lesser extent: it increases by no more than 2 times. Therefore, the relative content of the metal in the residual fraction (in silicates) decreases by more than 2 times (Tables 3.22, 3.23), which can be indicative of a technogenic source of metal input into the soil.

LEAD COMPOUNDS IN THE SOILS OF MONITORING PLOTS

Group of Weakly Bound Lead Compounds

The absolute and relative contents of exchangeable lead forms in the clean soil (plot 9) exceed the contents of lead complexes by 2.5 times (Tables 3.21, 3.22). The fractional composition of lead compounds is similar to that of copper compounds. In the contaminated soils, the most important changes occur for the complex compounds of the metal: their absolute content increases by more than 10 times; it makes up 5% of the total content and 15% of the content of weakly bound metal compounds. The relative content of exchangeable lead forms and those specifically bound to carbonates also slightly increases.

The difference in the composition of copper and lead compounds in the contaminated soils is that, in spite of the absolute increase in the lead content in Fe–Mn (hydr)oxides, the share of these compounds decreases with increasing contamination and becomes significantly lower than that of lead forms bound to carbonates. This tendency is observed in the soils of all monitoring plots.

Group of Strongly Bound Lead Compounds

The content of lead strongly bound to nonsilicate Fe and Mn compounds is a sensitive indicator of the effect of technogenic load and soil properties on the fixation of the metal. The soils of monitoring plots differ in the absolute content of this metal fraction by 5–14 times, which agrees with the differences in its shares in the strongly retained compounds by 3–8 times. The highest content of these lead compounds is observed in alluvial meadow soils rich in nonsilicate iron compounds. In the contaminated soils, the content of metals strongly retained by organic substances also increases. Under these conditions, the share of the metal in the residual fraction decreases.

Thus, organic substances are most active in the strong and weak retention of copper and lead in the soils of monitoring plots. Fe and Mn oxides and hydroxides also significantly contribute to the strong fixation of copper and lead in the contaminated soils.

ZINC COMPOUNDS IN THE SOILS OF MONITORING PLOTS

Group of Weakly Bound Zinc Compounds

The fractional composition of zinc compounds differs from those for copper and lead. In the clean soils, the content of weakly bound zinc compounds can be only 9%, among which 5% are specifically sorbed by carbonates and 2.6% by Fe and Mn (hydr)oxides (Table 3.23).

In the contaminated soils, the mobility of zinc increases more intensively than those of copper and lead. In soils of the plots close to the contamination source (Novocherkassk power station), the share of weakly bound zinc compounds exceeds one third of the total zinc compounds. Specifically bound forms are predominant among the weakly bound metal compounds.

The main distinction of the system of zinc compounds in the contaminated soils from the state of other metals is that an appreciable percentage of zinc (to 8% of its total content and 24% of its mobile compounds) occurs in the exchangeable form, i.e., in the least strongly retained state. In plot 5, the absolute content of readily exchangeable (extractable by 1 M $MgCl_2$) and difficultly exchangeable (extractable by 1 N AAB) metal forms increases by 23 times compared to the control soil (plot 9) (Table 3.21). In most cases, the content of exchangeable zinc compounds in a contaminated soil is no higher than the MCL, although the soils of the half of monitoring plots can be classified among contaminated soils according to their total content of zinc.

Soil organic matter little affects the formation of mobile zinc compounds. The content of complex zinc forms in soils is lower than those of the other metals; it is about 1 mg/kg (1–1.5% of the total metal content). This is related to the weak complexing capacity of zinc ions. Data are available that 1 g humic acids isolated from soils bound 0.99 meq Zn^{2+} and 2.39 meq Cu^{2+} at the same pH level, even if the amount of the Zn^{2+} ions added was double that of Cu^{2+} ions (Piccolo, Stevenson, 1982). Under these conditions, mobile zinc compounds in contaminated soil predominantly consist of exchangeable and specifically sorbed forms.

Group of Strongly Bound Zinc Compounds

Nonsilicate Fe and Mn compounds play the major role in the strong fixation of metal in the soils studied. They bind more Zn^{2+} ions than organic substances by 2–6 times. This fact agrees with the results of pot experiment (Section 3.2) and the data of other authors (Kosheleva et al., 2002; Ladonin, Plyaskina, 20030.

Thus, the mobilization and immobilization of zinc in technogenically contaminated soil are mainly related to the formation of compounds bound to Fe and Mn (hydr)oxides with different binding strength. The formation of mobile metal compounds is also favored by their participation in ion exchange.

The proportions of heavy metal compounds with different binding strengths to organic substances and Fe and Mn (hydr)oxides are given in Table 3.25. The increase in the content of weakly bound Cu, Zn, and Pb compounds with organic matter surpasses the increase in the content of their strongly bound forms. This is especially true for zinc, the strong fixation of which involves little organic matter.

Table 3.25. Table 3.14 The ratio of weakly bound /strongly bound compounds of Cu, Pb and Zn with organic matter and Fe and Mn (hydr)oxides, % of their total contents within this components

№ of the plot	Organic matter			Fe and Mn (hydr)oxides		
	Cu	Pb	Zn	Cu	Pb	Zn
1	20/80	36/64	40/60	31/69	32/68	57/43
2	39/61	12/88	29/71	48/52	9/91	30/70
3	11/89	12/88	34/66	32/68	19/81	40/60
4	23/77	20/80	45/55	38/62	29/71	46/54
5	28/72	27/73	31/69	34/66	39/61	42/58
6	22/78	19/81	21/79	32/68	31/69	42/58
7	15/85	14/86	14/86	41/59	27/73	46/54
8	31/69	19/81	35/65	57/43	34/66	41/59
9	10/90	7/93	17/83	29/71	54/46	29/71
10	18/83	29/71	16/84	25/75	38/62	25/75

Thus, aerosol emissions from the Novocherkassk power station are the major agent of technogenic impact on the soils of Rostov oblast; vehicle exhausts can be a source of additional lead emission. The soils occurring within a radius of 5 km from the Novocherkassk power station along the predominant wind direction are subjected to the highest contamination. The contents of Cu, Zn, and Pb in soils of these areas exceed their MCL levels.

In contaminated soils, the proportions of metal compounds change. In distinction from the clean soils, where the metal strongly retained in the structure of silicate minerals were predominant (48–78% of the total content), the share metals in the minerals of contaminated soils increases, as well as the content of their weakly bound compounds. These changes are proportional to the load of heavy metals.

The soils of monitoring plots differ in the capacity of resisting to the increase in metal mobility under contamination. The lower the soil buffering for metals, the higher their ecological hazard. An ecologically hazardous level of the most mobile exchangeable metal forms is developed in the contaminated soils with lower buffering capacities. According to the ecological hazard caused by the contamination with heavy metals, the soils near the Novocherkassk power station form the following increasing series: sandy clay meadow-chernozemic soil < clay loamy meadow-chernozemic soil < clay loamy ordinary chernozem < sandy alluvial meadow soil.

The group composition of different metal compounds varies among the contaminated soils. Copper and lead arrived with technogenic emissions remain in the mobile state predominantly as organomineral complexes, and zinc remains in the exchangeable form and in the form of compounds specifically sorbed by Fe and Mn (hydr)oxides. Organic substances and Fe and Mn nonsilicate minerals make the largest contribution to the strong fixation of copper and lead; nonsilicate Fe and Mn minerals mostly contribute to the strong fixation of zinc.

3.2.4. Information Value of the Group Composition Parameters of Metal Compounds in Soils

The testing of methods for determining the group composition of metal compounds in soils under different conditions (laboratory, pot, field, and production experiments) confirm the validity of the developed system and the information value of parameters determined using this system for assessing the ecological status of soils.

The system of methods for determining the group composition of metal compounds in soils includes the following operations:

(1) The simultaneous use of parallel and sequential extractions for the fractionation of metal compounds in soils.

(2) The identification of separate heavy metal compounds differently bound by soil components (nonsilicate Fe and Mn compounds, organic matter, carbonates) by calculation.
(3) The classification of metal compounds by their binding strength to soil components (strongly and weakly bound) and the calculation of metal mobility in the soil from the ratio between the metal contents in two groups of metal compounds.
(4) The calculation of the contributions of different fractions to the metal group composition and the assessment of their role in the changes in metal mobility.

The similarity of the results of metal fractionation obtained using parallel extractions and the combined fractionation scheme indicates that both fractionation procedures can be used for ecological purposes (Tables 3.26, 3.27).

Table 3.26. Comparison of results of determining Cu, Pb, and Zn compounds in the soils of pot experiment using (1) parallel extractions and (2) the combined fractionation scheme

Dose of metal, mg/kg	Weakly bound /strongly bound compounds					
	Cu		Pb		Zn	
	1	2	1	2	1	2
Control (no added metal)	5/95*	5/95	14/76	12/88	11/89	10/90
100	28/72	26/74	26/74	25/75	47/53	39/61
300	33/77	30/70	3278	31/69	37/63	40/60

The method of parallel extractions can be used for the rapid assessment of heavy metal mobility in the monitoring observations of soils. The combined fractionation system is efficient in studying the transformation of metals in contaminated soils, because it can not only separate the metal compounds with different mobilities under specific conditions, but also predict the behavior of pollutants.

The assessment of the status of heavy metals in the clean soils of Rostov oblast and its changes under contamination allowed revealing general tendencies manifested under conditions of model, pot, and field experiments and monitoring observations (Tables 3.28, 3.29). It was shown that strongly bound compounds of heavy metals prevail in the original soils, mainly in silicates. Among mobile metal compounds, those specifically sorbed on carbonates are predominant. At the additional input of heavy metals into the soil (especially from a permanent contamination source), the equilibrium in the system of metal compounds is shifted toward an increase in metal mobilities.

All soil components are responsible for the retention of metals in both mobile and strongly bound states. Carbonates and silicate minerals play the leading role in the mobilization and immobilization of natural metal compounds in ordinary chernozem; for exogenic metal compounds, these are organic matter and Fe and Mn (hydr)oxides.

Table 3.27. Comparison of results of determining Cu, Pb, and Zn compounds in the soils of monitoring plots using (1) parallel extractions and (2) the combined fractionation scheme

№ of the plot	Weakly bound /strongly bound compounds						
	Cu		Pb		Zn	
	1	2	1	2	1	2
1	28/72	23/77	37/63	26/74	41/59	32/68
2	25/75	24/76	22/78	20/80	38/62	27/73
3	18/82	14/86	24/76	17/83	29/71	25/75
4	32/68	25/75	33/67	28/72	40/60	35/65
5	36/64	28/72	41/59	33/67	42/58	32/68
6	35/65	28/72	35/65	28/72	39/61	31/69
7	26/74	19/81	39/61	25/75	26/74	28/72
8	33/67	26/74	26/74	24/76	35/65	29/71
9	19/81	18/82	20/80	20/80	13/87	14/86
10	13/87	11/89	32/68	26/74	11/89	9/91

The determination of the group composition of metal compounds also revealed the peculiar features of Pb, Cu, and Zn compounds in soils and the effect of soil properties. Organic substances and iron–manganese (hydr)oxides have the leading effect on the transformation of Pb and Cu compounds in contaminated soils; for Zn compounds, these are iron–manganese (hydr)oxides.

The content of metals in the residual fraction bound to silicates is more stable and less subjected to changes compared to other fractions; therefore, the relative content of heavy metals in this fraction can serve as an indicator of the source (natural or anthropogenic one) of metal accumulation in soils. For example, a decrease in the share of metals in this fraction by 20–30% was observed at the contamination of soils in pot and field experiments and in monitoring plots.

Table 3.28. Comparative analysis of the group composition and mobility parameter (Km) of Zn and Pb compounds in the soils of pot and field experiments during two years after metal addition

Dose of metal, mg/kg	WB/SB, % of the total content					Exchangeable/ complex/ specifically sorbed, % of the content of weakly bound metals					Km				
	pot experiment		field experiment		pot experiment		field experiment		pot experiment		field experiment	
	1 year	2 year	1 year	2 year	1 year	2 year	1 year	2 year	1 year	2 year	1 year	2 year
Pb												
Control (no added metal)	12/88	14/86	15/85	15/85	18/9/73	19/9/72	23/9/68	24/8/68	0,2	0,2	0,2	0,2
100	28/72	26/74	38/62	45/55	23/16/61	14/29/57	31/14/55	24/28/48	0,4	0,4	0,6	0,8
Zn												
Control (no added metal)	10/90	11/89	12/88	12/88	7/1/92	5/4/91	8/5/87	8/6/86	0,1	0,1	0,1	0,1
300	31/69	37/63	32/68	35/65	23/26/51	17/15/68	29/25/46	23/19/58	0,5	0,6	0,5	0,5

Pot and field experiments showed that the proportions of metal compounds in soils varied during two years after contamination: the share of specifically sorbed zinc forms in the mobile zinc compounds increased, as well as the share of complex copper and lead forms in their mobile compounds (Table 3.28).

Table 3.29. Proportions of Cu, Pb, and Zn compounds in ordinary chernozem under single and permanent polymetallic contamination

The total content, mg/kg	Weakly bound / strongly bound compounds	
	pot experiment	monitoring plots
Cu		
40-43	12/88	26/74
52	19/81	28/72
100 in the experiment, 75 in plots	26/74	33/67
Pb		
25	12/88	20/80
30	17/83	38/62
37-45	16/84	32/68
60-64	25/75	41/59
Zn		
67-70	10/90	11/89
90	23/77	26/74
107-116	41/59	40/60

It was found that soil properties affect the mobility of metals. A higher mobility of Cu, Zn, and Pb was noted in the sandy alluvial meadow soils of monitoring plots compared to the adjacent sandy clay meadow-chernozemic soils. A higher role of exchange processes in the transformation of heavy metals was revealed in chestnut soil compared to chernozem.

The duration of technogenic impact affects the mobility of pollutants in soils. At the similar contents of metals in chernozem, the long-term (for more than 40 years) influx of pollutants to the soil results in a more significant increase in their mobility than a single exposure (Table 3.29).

Chapter 4

SORPTION AND DISTRIBUTION OF HEAVY METALS IN CONTAMINATED SOILS

Information on the capacity of soils to adsorb heavy metals under contamination can be derived not only from the analysis of technogenically disturbed soils or those artificially contaminated with metals under conditions of field or pot experiments. Such information can also be acquired in laboratory experiments on the sorption of metals by soils. Parameters of the sorption capacity of soils are actively used for assessing the actual and potential capacities of soils to sorb metals (Pinsky, 1997; Ladonin, 2000; Ponizovskii, Mironenko, 2001; Vodyanitskii, 2005b, Minkina et al., 2005). However, the identification of metal compounds in soils is also of importance. For this purpose, it is advisable to simultaneously use the parameters of the sorption capacity of soils for metals and the group composition of metals sorbed. The integrated study of the sorption of metals by soils and the formation of different compounds by the metals sorbed contributes to the understanding of transformation mechanisms of metal compounds in the soil and the assessment of tendencies in changes of proportions of weakly and strongly bound soils under contamination conditions.

4.1. APPROACHES TO REVEALING RELATIONSHIPS BETWEEN THE SORPTION CAPACITY OF SOILS FOR METALS AND THE FORMATION OF THEIR COMPOUNDS IN SOILS

Sorption is the uptake of a substance (sorbate) by the solid phases (sorbent) of a heterogeneous system, which results in the removal of the substance from the solution (or the air environment) and its accumulation on the solid phase surface. Uptake is a component process of interaction between heavy metals and soil components. This process is simulated in laboratory experiments under static and dynamic conditions in order to rapidly determine the sorption parameters of heavy metals (Pinsky, 1997; Ladonin, 2000; Ponizovskii, Mironenko, 2001; Vodyanitskii, 2005a; Minkina et al., 2005).

The capacity of soils to sorb metals is described by sorption isotherms. They show the amount of a metal sorbed by the soil (mg/kg, mmol/kg) as a function of its concentration in the solution (mg/l, mmol/l). The relationship between these parameters is described by different equations. The empirical Langmuir equation is the most frequently used. It has the form

$$X/m = S_{max}\, kC/(1+kC),$$

where X/m is the amount of metal ions sorbed per unit mass of sorbent; C is the equilibrium concentration of metal in the contacting solution; S_{max} is the maximum adsorption of metal ions; and k is the equilibrium constant.

The relationship between these parameters is nonlinear when considered over a wide range of metal concentrations in the solution.

The Langmuir equation can be transformed into a linear form, which relates the reciprocal values of the adsorbed metal amount and the equilibrium metal concentration in the solution:

$$m/X = C/S_{max} + 1/k\, S_{max} \times 1/C$$

By solving this equation, the maximum sorption capacity of the sorbent (S_{max}) can be calculated, as well as the ability of a substance to be adsorbed and the strength of its retention by the sorbent (K).

The nonlinear character of the relationship between the amount of substance sorbed and its concentration in the solution indicates that the adsorption of metal by the soil decreases with increasing its concentration in the solution. This is attributed to the change in mechanisms of metal retention by the sorbent and the successive involvement of new sorption centers in the process. It is suggested that these centers are polyfunctional in nature, i.e., that the same soil components (e.g., organic substances or Fe and Mn hydroxides) are capable to fix metals strongly and weakly.

In this work, the determination of Cu, Pb, and Zn contents in such compounds (using the combined scheme for their fractionation) was combined with the determination of the sorption capacity of soils for these elements. The study included five successive stages.

The first stage involved standard experiments on the sorption of metals by soils; isotherms for metal sorption in soils were then plotted; their regions where the linearity was disturbed were found, and the corresponding metal concentrations in the solution interacting with the soil were determined.

At the second stage, soil samples were brought into contact with solutions with increasing metal concentrations (selected at the first stage).

At the third stage, the content of metal compounds in thus contaminated soil samples was determined in accordance with the selected fractionation scheme.

At the fourth stage, an isotherm of metal sorption was plotted for each isolated metal compound, which reflected changes in this metal form as a function of its concentration in the contacting solution.

At the fifth stage, the maximum sorption and binding strength of metal ion were calculated for each compound from the Langmuir equation.

The obtained values of maximum sorption were considered as indicators for the probability of inclusion of metal ions in specific ionic compounds and compared with the actual contents of these compounds in the soil.

4.2. PROCEDURE OF MODEL LABORATORY EXPERIMENTS

Experiments were conducted with samples of ordinary chernozem, whose properties are given in Table 3.1.

Experiments on metal sorption were carried out under static conditions. Soil samples were triturated to <1 mm and treated with solutions of Cu, Zn, and Pb acetates with the following concentrations (mmol/l): 0.05, 0.08, 0.1, 0.3, 0.5, 0.8, and 1.0 (at a soil : solution ratio of 1 : 10). The suspensions were shaken for 1 h and left to stand for 24 h. Then, the solutions were filtered, and the concentrations of metals in the filtrates were determined by atomic absorption spectroscopy. The content of ions sorbed was calculated from the difference between the metal concentrations in the initial and equilibrium solutions. Experiments were performed in triplicate.

4.3. RESULTS OF LABORATORY SIMULATION

From the results of laboratory experiments, Cu, Pb, and Zn sorption isotherms were plotted (the concentration of metal in the equilibrium solution was used as the abscissa, and the content of metal sorbed by the soil was used as the ordinate) (Fig. 4.1). It was found that lead and copper were most actively sorbed by chernozem. At a metal concentration of 0.02 mmol/l, the amounts of lead and copper ions retained by chernozem samples were larger than those of zinc ions by almost 5 times.

Four linear segments were found in each isotherm of Cu, Pb, and Zn sorption by the soil. They corresponded to the following metal concentrations in the contacting solutions (mmol/l): 0.05, 0.5, 0.8, and 1.0. After the solutions were separated from the solid phases, 4 contaminated soil samples were obtained for each metal: Cu, Pb, and Zn. In all samples, the contents of metals in 7 fractions were determined using the combined fractionation scheme: weakly bound compounds (exchangeable, complex, specifically sorbed by carbonates, specifically sorbed by Fe and Mn (hydr)oxides) and strongly bound to organic substances, Fe and Mn (hydr)oxides, and silicates (Table 4.1).

The group composition showed that strongly bound metal forms prevailed in the solid phases only at the low concentration of metals in the solution (metal application rate 0.5 mmol/kg). At a higher load, the content of weakly bound compounds became comparable to that of strongly bound compounds. This fact agreed with the tendencies of changes in the proportions of metal compounds in contaminated soils revealed in pot experiments (Tables 4.2, 4.3). At a very high Pb application rate (8–10 mmol/kg), weakly bound metal compounds became predominant.

Noteworthy changes in the group composition of metals compounds were observed under contamination conditions (Table 4.3). The relative contents of exchangeable and complex compounds of three metals increased with increasing load. The relative contents of metals strongly retained by silicates stably decreased. In the strongly bound metal compounds, the predominance passed from the silicate-bound compounds to those fixed by organic substances for Pb and Cu and to the compounds retained by Fe and Mn oxides and hydroxides for Zn. The same soil components increased their activity in the weak retention of metals.

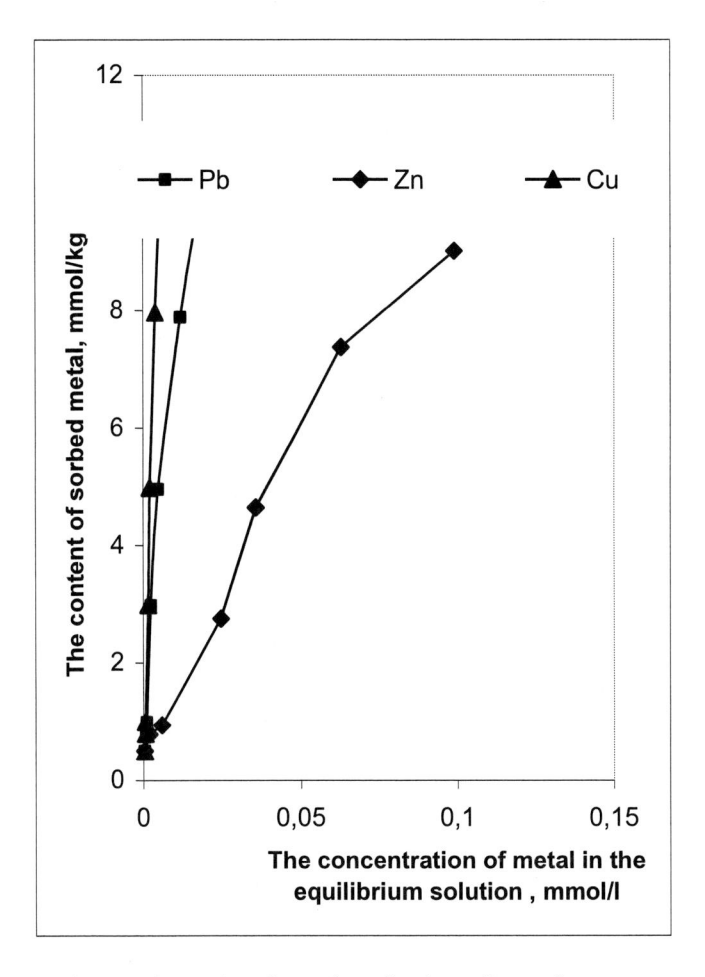

Figure 4.1. Isotherms of Cu, Pb and Zn adsorption by ordinary chernozem.

Carbonates lost their predominance in the weak retention of metals under increasing metal load. The tendencies revealed in the analysis of soils contaminated with metals under laboratory conditions confirmed the results obtained in the pot and field experiments.

Table 4.1. The content of Cu, Pb and Zn adsorbed by chernozem and their distribution among groups of compounds, mM/kg

Dose of metal, mM/kg	Weakly bound compounds (WB)			Specifically sorbed		Strongly bound compounds (SB)			Sum of fractions
	Exchangeable AAB/MgCl$_2$	Complex	on carbonates	on on Fe and Mn (hydr)oxides	With organic matter	With on Fe and Mn (hydr)oxides	With silicates	
Cu								
0,5	0,02/---	0,02	0,07	0,009	0,29	0,05	0,72	1,16
5	0,2/0,12	0,8	0,65	0,17	1,79	1,12	0,73	5,38
8	0,6/0,36	1,8	0,96	0,65	2,89	1,9	0,9	9,46
10	0,76/0,43	2,17	1,06	0,81	3,32	2,25	0,96	11,00
Pb								
0,5	0,03/0,002	0,036	0,09	0,04	0,16	0,098	0,11	0,54
5	0,35/0,33	0,86	0,57	0,34	1,69	1,04	0,23	5,06
8	0,86/0,78	2,16	1,08	0,53	2,6	1,54	0,25	8,94
10	1,90/1,70	2,96	1,21	0,61	2,8	1,79	0,27	10,30
Zn								
0,5	0,04/0,02	0,03	0,13	0,02	0,05	0,35	0,76	1,36
5	0,57/0,31	0,51	1,0	0,69	0,3	2,3	1,0	6,11
8	0,89/0,72	0,89	1,3	1,42	0,48	3,5	1,09	9,40
10	1,08/0,91	1,05	1,45	1,71	0,52	3,9	1,13	10,67

Table 4.2. Relative contents of Cu, Pb and Zn compounds in polluted chernozem, % of the total content

Dose of metal, mM/kg	Weakly bound compounds (WB)		Specifically sorbed		Strongly bound compounds (SB)			WB/SB
	Exchangeable, MgCl₂	Complex	on carbonates	on Fe and Mn (hydr)oxides	With organic matter	With on Fe and Mn (hydr)oxides	With silicates	
Cu								
0,5	0	1,7	6,0	0,8	25,0	4,3	62,1	9/91
5	2,2	14,9	12,1	3,2	33,3	20,8	13,6	32/68
8	3,8	19,0	10,1	6,9	30,5	20,1	9,5	40/60
10	3,9	19,7	9,6	7,4	30,2	20,5	8,7	41/59
Pb								
0,5	0,4	6,7	16,8	7,5	29,9	18,3	20,5	31/69
5	6,5	17,0	11,3	6,7	33,4	20,6	4,5	42/58
8	8,7	24,2	12,1	5,9	29,1	17,2	2,8	51/49
10	15,0	26,1	10,7	5,4	24,7	15,8	2,4	57/43
Zn								
0,5	1,5	2,2	9,6	1,5	3,7	25,7	55,9	15/85
5	5,1	8,3	16,4	11,3	4,9	37,6	16,4	41/59
8	7,7	9,5	13,8	15,1	5,1	37,2	11,6	46/54
10	8,5	9,8	13,6	16,0	4,9	36,6	10,6	48/52

Table 4.3. Relative contents of Cu, Pb and Zn forms in groups of weakly and strongly bound compounds

Dose of metal, mM/kg	Weakly bound compounds (WB)			Strongly bound compounds (SB)			
	Exchangeable, MgCl$_2$	Complex	Specifically sorbed		With organic matter	With on Fe and Mn (hydr)oxides	With silicates
			on carbonates	on on Fe and Mn (hydr)oxides			
Cu							
0,5	0	20	71	9	27	5	68
5	7	46	37	10	49	31	20
8	10	48	25	17	51	33	16
10	10	49	24	18	51	34	15
Pb							
0,5	1	21	54	24	43	27	30
5	16	41	27	16	57	35	8
8	17	47	24	12	59	35	6
10	26	46	19	9	58	37	6
Zn							
0,5	10	15	65	10	4	30	66
5	12	20	40	27	8	64	28
8	17	21	30	33	9	69	21
10	18	21	28	33	9	70	20

The absolute content of all strongly and weakly bound metal compounds in the soil increased with increasing metal load. Therefore, the content of metals retained by the soil in specific compounds as a function of metal concentrations in the contacting solution could be plotted for each of 7 separated metal fractions.

The curves plotted were satisfactorily described by the Langmuir equation for heterogeneous exchangers. Adsorption parameters were determined by the approximation of accumulation curves for each metal form (Table 4.4). The maximum adsorption value, S_{max}, corresponded to the largest amount of metal capable to bind to the surface due to the formation of the presumed compound. The affinity of metal for the adsorption center and the strength of the formed bond were estimated by the value of constant K.

The calculated parameters were characterized by a significant coefficient of determination (R^2). The potential capacity of soil for sorbing Cu, Zn, and Pb ions in the exchangeable form was high, but it was significantly lower than the CEC of the soil studied. This form of metals was weakly retained by soil components and could rapidly transform into more strongly compounds. Therefore, the actual content of exchangeable metal compounds in the soil was very low. This was also true for the metal compounds weakly retained by organic substances in the form of complexes. The lowest values of the maximum retaining capacity for Cu, Zn, and Pb were observed in silicate minerals; however, they fixed the metals most strongly.

The parameter of maximum metal sorption corresponded only to the content of the most strongly fixed metal compounds, namely in silicates. For all other metal compounds, their actual contents in clean and contaminated soils were far from the maximum capacity of soils to sorb metals in these compounds (Table 4.5). This was related to the fact that the maximum capacity of soils to sorb and retain metals in these compounds could not be reached at the permanent transformation of these compounds under dynamic equilibrium conditions.

The thermodynamic assessment of heavy metal groups in the soil was performed using the integrated analysis of sorption isotherms and sorbed metal forms. It was found that these metal groups essentially differ in sorption parameters, especially in the binding strength (Table 4.6).

Table 4.4. Thermodynamic parameters of Cu, Pb, and Zn compounds in the soil

Metal compounds	Cu S_{max}, мМ/кг	Pb k	Zn R^2	S_{max}, мМ/кг	k	R^2	S_{max}, мМ/кг	k	R^2
Exchangeable	74,3±32,6/ 22,4±10,5	1,9±0,6/ 1,6±0,7	0,95/ 0,91	2,6±1,0/ 2,8±1,1	37,3±13,0/ 29,7±10,1	0,99/ 0,99	6,1±0,3/ 5,2±1,0	10,9±3,0/ 2,2±0,5	0,99/ 0,97
Complex	42,3±16,7	9,98±4,0	0,97	8,6±4,0	35,6±4,1	0,99	2,3±0,7	8,9±5,0	0,99
Specifically sorbed on carbonates	2,4±1,1	154,9±30,1	0,97	2,0±0,3	84,5±7,9	0,99	1,93±0,3	31,3±12,0	0,99
Specifically sorbed on on Fe and Mn (hydr)oxides	63,0±10,2	2,3±0,9	0,93	0,9±0,1	128,0±8,0	0,99	5,4±3,8	4,9±3,7	0,98
Strongly bound with on Fe and Mn (hydr)oxides	9,7±4,3	56,7±22,2	0,98	2,5±0,1	141,1±12,8	0,99	6,26±2,8	77,9±12,1	0,97
Strongly bound with organic matter	9,2±3,8	107,9±28,1	0,99	3,9±0,3	158,8±7,7	0,99	0,85±0,3	17,1±6,2	0,97
Silicates	0,92±0,1	5536,0±382,0	0,64	0,27±0,2	1470,0±243,0	0,99	1,1±0,04	4759±335,5	0,90

Table 4.5 Cu, Pb, and Zn compounds determined by fractionation in the soils of pot experiment (mg/kg) and the maximum capacity calculated for them from sorption isotherms (S_{max})

Metal compounds	The content of Cu compounds in soil			The content of Pb compounds in soil			The content of Zn compounds in soil		
	no added metal	Doze of metal, 300 mg/kg	S max	no added metal	Doze of metal, 300 mg/kg	S max	no added metal	Doze of metal, 300 mg/kg	S max
Weakly bound									
Exchangeable	0,3	14	4726	0,6	12	539	0,4	23	397
Complex	0,2	45	2690	0,3	46	1782	0,3	21	150
Specifically sorbed on carbonates	1,7	37	153	1,6	33	414	6,3	51	126
Specifically sorbed on on Fe and Mn (hydr)oxides	0,2	19	4007	0,7	12	187	0,3	40	351
Strongly bound									
with on Fe and Mn (hydr)oxides	4	107	585	7	114	808	1	4	55
with organic matter	1	76	617	2	75	518	6	105	407
with silicates	36,9	69	59	14	37	56	56	84	72

The group of weakly bound Cu, Pb, and Zn compounds was characterized by a low binding strength of metals to soil and a high sorption capacity. The group of strongly bound compounds of these elements had a high binding strength and a low sorption capacity. The lowest binding strength in the group of strongly bound compounds was found for zinc.

Table 4.6. Thermodynamic parameters of Cu, Pb, and Zn weakly bound (WB) and strongly bound (SB) compounds in chernozem

Group of metal	S_{max}, мМ/кг	k	R^2
Cu			
WB	86,0±33,8	10,3±4,1	0,97
SB	14,2±1,7	159,5±31,2	0,99
Pb			
WB	12,3±1,08	45,6±16,5	0,99
SB	6,39±0,64	179,3±54,0	0,99
Zn			
WB	11,0±3,3	9,1±4,6	0,99
SB	7,7±3,0	26,8±10,5	0,90

Thus, although the separation of metal compounds based on sorption capacity parameters is rather conventional, this approach allows the metal compounds to be classified in accordance with the binding strength to soil components. The results show a general tendency: the compounds of heavy metals formed during their sorption by the soil are less stable and, hence, more mobile than their natural compounds, which determines the ecological hazard of metal-contaminated soils.

The parameters of metal binding strength by different soil components are essential for ensuring the protection properties of an ecosystem.

Chapter 5

TRANSFORMATION OF HEAVY METAL COMPOUNDS DURING THE REMEDIATION OF CONTAMINATED SOILS

The barrier function is an essential function of soil in an ecosystem. The soil protects natural waters, air, and plants from pollutants. This function is ensured by the sorption capacity of soils. It should be noted that plants also have some resistance to soil contamination.

The soil under study is a highly buffering self-protecting system. However, our studies showed that the self-protection mechanisms are not always efficient. Therefore, methods should be developed for the remediation of polluted soils with the use of ameliorants.

The available data on the efficiency of specific ameliorants are insufficient to reveal the processes resulting in the redistribution of heavy metal forms in contaminated soils.

In addition, the efficiency of remediation methods was better studied for slightly acid chernozems and acid soddy-podzolic and podzolic soils with different particle-size distributions. At the same time, ordinary chernozems usually escape the attention of researchers.

The soils of this genetic subtype occupy a significant area (mainly in the industrial regions of Rostov oblast and Krasnodar region subjected to active anthropogenic pollution) and their major part is in agricultural use; therefore, analogous studies of these soils are of current interest.

5.1. SUBSTANTIATION OF AMELIORANT SELECTION FOR THE REMEDIATION OF CHERNOZEM

The reclamation methods of arable soils contaminated with metals are mainly aimed at decreasing the content of mobile metal compounds. The selection of ameliorants should be based on the mechanisms of strong metal fixation; i.e., the action of an ameliorant should be directed to the enhancement of the barrier function of soils. Insufficient data are available on mechanisms for the strong fixation of metals by soil components, including in reclaimed soils.

The analysis of the group metal composition showed the polyfunctional nature of soil components and the capacity of each of them for both strong and weak retention of contaminant metals. The selection of ameliorants for contaminated soils was based on the presumed formation of metal compounds strongly bound to soil components corresponding to their weakly bound analogues.

The strong fixation of heavy metals in the soil is due to their chelation, precipitation, and fixation in the structure of minerals; therefore, manure (active agent in the complexation of metals with differently stability), chalk (active agent in the specific sorption and precipitation of metals), and glauconite (active agent in the exchangeable sorption and fixation of metals) were used as ameliorants.

The high capacity of humus acids for binding and strongly retaining appreciable amounts of heavy metals found practical application for soil detoxification (Provisional Recommendations…, 1990; Perez de Mora et al., 2003; Choi et al., 2008). Brown coal, which is also a source of natural humic acids, is used as a sorbent (Bezuglova et al., 1996).

Natural zeolites having a high sorption capacity have attracted recent attention of scientists dealing with the protection of soils and plants from the contamination with heavy metals (Mineev et al., 1989; Ming, Mumpton, 1989; Distanov, Konyukhova, 1990; Baidina, 1994; Knox, Adriano, 1999; Barbu et al., 2003; Kliaugiene, Baltrenas, 2003). Natural sorbents are profitable agents, because they are ecologically safe, readily available, and inexpensive. There are different opinions about the efficiency of zeolites on the soils contaminated with heavy metals. Some authors (Ponizovskii et al., 2003; Belousov, 2006) emphasized their high selectivity for heavy metals. Other authors (Baidina, 1994; Dabakhov et al., 1998) reported the lower efficiency of zeolites compared to lime materials and a significant limitation in the mobility of heavy metals in the soil and their input into plants only at zeolites application rates of 80–100 t/ha and more. This is also true for the effect of organic fertilizers (Dabakhov et al., 1998; Chernykh et al., 1995; Lebedeva et al., 1997).

At the same time, it should be noted that the use of zeolites as sorbents of heavy metals is limited by the following factors. First, the volume of zeolites applied is very large, which makes them applicable only near zeolites fields. Second, along with heavy metal cations, zeolites can sorb potassium, ammonium, and microelement ions, i.e., affect the conditions of the mineral nutrition of plants (Baidina, 1994). Third, data are available that zeolites are subjected to weathering, during whish they can be transformed into other minerals with different properties of cation sorption (Il'in, 1991). Fourth, exchange on some zeolites minerals proceeds even more slowly than on clay minerals. The completion of this process, i.e., the penetration of exchangeable cations into channel-shaped holes, takes much time (Grin, 1959). This property of zeolites is used in plant growing: they are mixed with organic fertilizers (Loboda, 2000). Long-acting composite fertilizers are prepared by this method.

Glauconite fields are common in Rostov oblast. Glauconite enters in sands, sandstones, clay, and marls (Katsnel'son, 1981). Total predicted resources of the mineral in Rostov oblast exceed 20 million m^3 (Khardikov et al., 1999).

The adsorption properties of zeolites are determined by the unique crystal lattice characterized by a developed internal surface and a strictly determined size of input windows. Zeolites are a sort of molecular sieves capable to sorb molecules of specific size from their mixtures. Only molecules whose size is smaller than the input window can penetrate into the adsorption cavity (Glazkova et al., 2003). Zeolites can adsorb relatively large amounts of heavy metal salts. The CEC of natural zeolites is 100–300 mg/kg (Pinsky, 1997).

A large number of publications deal with the use of natural zeolites for the purification of water from dissolved chemical impurities (Bingham, Page, 1964; Hildebrand, Blum, 1974; Glazkova et al., 2003; Gerasimova, 2003), but the use of zeolites as adsorbents of heavy metals is still insufficiently understood (Panin, Bairova, 2005).

A higher efficiency of the simultaneous application of ameliorants compared to their separate application was noted repeatedly (Baidina, 1994; Chernykh et al., 1995; Chen, Looi, Liu, 1999; Geebelen, Nangronsveld, Clijsters, 1999). In the recommendations for decreasing the toxicity of soils contaminated with heavy metals (Determination of Toxicity…, 1985), it was proposed to use lime materials and high rates of organic fertilizers (100–150 t/ha).

There are also many less common methods. The claying of coarse soils, which significantly increases their cation exchange capacity, can give good results (Sizov et al., 1990). An expensive method for decreasing the mobility of heavy metals involves the use of ion-exchange resins containing carboxyl and hydroxyl groups. Resins are used in the acid form or saturated with potassium, calcium, or magnesium ions or their mixture and applied to the soil as granules or powders (Minkina et al., 2003).

Many authors noted the high efficiency of liming the contaminated soils (Baidina, 1994; Chernykh et al., 1995; Basta, Armstrong, Hanke, 2001; Brown et al., 2001; Alekseev, Byalushkina, 2002; Yang et al., 2008). The Rostov oblast has immense reserves of limestone (Reference Book..., 2000). However, liming is widely used for acid soils but not for neutral and slightly alkaline soils. Therefore, the task was set to determine the group composition and mobility of heavy metals in soils with different contents of carbonates for studying the possibility of using chalk deposits in ordinary chernozems contaminated with Cu, Zn, and Pb.

5.2. INTERACTION MECHANISMS OF HEAVY METALS WITH CARBONATES IN SOILS

Carbonates are especially important in calcareous chernozems, which have a one-layered alkaline vertical profile, in distinction from other chernozemic soils. The carbonate content is not a formal morphological feature; it reflects the specificity of aerohydrothermal conditions in these soils and the formation conditions of mobile chemical compounds (Krupenikov, 1979).

The Lower Don basin, along with the Central and Western Ciscaucasia, is a classical region of calcareous ordinary chernozems. The content of $CaCO_3$ in the upper (0- to 20-cm) layer of these soils varies in a wide range, from 0.1 to 4.1% on average. Its content gradually increases down the profile and reaches 6–19% in the horizons of accumulation of carbonate nodules (Krupenikov, 1979).

Ordinary chernozems are characterized by specific distributions of free carbonates. Varied amounts of free carbonates occur at different depths of the 2-m thick soil layer. In these soils, pH is 6.5–8.0 in the upper horizons and 7.5–8.7 in the lower horizons (Val'kov, 1977).

Another characteristic feature of ordinary chernozems is the presence of micellar carbonates in the form of carbonate mold along with common carbonate neoformations (beloglazka and veins) (Val'kov, 1977; Bezuglova, 2001; Val'kov, 2002). Permanent positive temperatures in the chernozem profiles of the South-European facies affect the high migratory capacity of soil solutions with the high content of calcium bicarbonates and the formation of $CaCO_3$ neoformations of migratory (micellar) type (Minkin, 1974). Ordinary chernozems are frequently called micellar-calcareous chernozems. Mild winters, slight winter frosts, deep soil wetting, lasting warm periods, and the alternation of downward and upward moisture fluxes result in a significant variation of carbonate migration along the soil profile and favor the development of micellar neoformations clearly observed on drying soil sections.

Carbonate mycelium is clearly defined at a depth of 20–30 cm below the effervescence line to the appearance level of $CaCO_3$ nodules (Gavrilyuk et al., 1976). It consists of the most mobile carbonate material, ikaite (needle calcite) $CaCO_3 \cdot 6H_2O$ (Krupenikov, 1979). This mineral is predominant in the upper horizons of chernozems developed on loesses; calcite and dolomites also appear with depth.

Calcite is present in all particle-size fractions; dolomite is concentrated in the fraction 0.1–0.01 mm. The highest contents of carbonates are present in the fraction 0.01–0.001 mm, but their major part (40–60%) is found in the clay fraction. The redistribution of carbonates among particle-size fractions occurs in the upper calcareous horizons: the content of coarsely dispersed carbonates decreases and that of finely dispersed carbonates increases.

The presence of carbonates in the soil ensures the calcium–carbonate equilibrium, which affects the chemical composition of soil solution and the dissolution, migration, and precipitation of carbonates in the soil profile, as well as the character of ion-exchange processes at the interface (Minkin et al., 1995).

The content of carbonates in the soil and the pH level are closely related and affect the transformation of heavy metal compounds in soils (Papadopoulos, Roweli, 1988). Carbonates are capable chemically sorb metals on their surface. The effect of pH is manifested in the direct competition of protons with metals for adsorption sites and through changes in the forms of metal ions in the solution and the solid phases to which they are bound (Gorbatov, Zyrin, 1988).

The effect of carbonates on the group composition of copper, zinc, and lead compounds was studied using clean soils containing 0.5% carbonates from the pot experiment and contaminated soils with similar carbonate contents to which were separately added Cu, Zn, and Pb acetates at a rate of 300 mg/kg. In additional treatments, CaCO$_3$ was added at rates of 2.5 and 5% and the metals studied were separately added at 300 mg/kg.

It was shown that carbonates significantly affect the composition of metal compounds. They also affect the group and fractional composition of Cu, Zn, and Pb compounds in contaminated and clean soils.

The group of weakly bound compounds in both clean and contaminated soils mainly consists of the metals specifically sorbed by carbonate minerals.

When the carbonate content in soils increases, the mobility of metals decreases and the proportions of weakly bound metal compounds change: the absolute and relative contents of the most mobile exchangeable and complex forms decreases and the content of metal compounds specifically sorbed by carbonates increases (Tables 5.1, 5.2).

At a carbonate content of 5%, the share of Pb and Cu compounds specifically sorbed by carbonates is 84–87%, and that of the corresponding Zn compounds is 100% (Table 5.3). The combined fractionation scheme does not involve the determination of metal compounds strongly retained by carbonates. Nonetheless, the existence of such forms is not denied. Under the fractionation conditions, they could be classified among the metals compounds strongly bound to Fe–Mn (hydr)oxides.

Yu. N. Vodyanitskii (2005b) also made a similar supposition and noted that lead compounds retained by carbonates can be partly identified as bound by Fe and Mn oxides in the fractionation using the Tessier procedure. The content of these metal compounds in soils increases with increasing carbonate content.

Thus, the low mobility of metals in combination with the high share of weakly bound heavy metals specifically sorbed by carbonates can be considered as a regional feature of the soils studied. The low content of mobile metal compounds in ordinary chernozem was previously attributed to their fixation by finely dispersed micellar carbonates under slightly alkaline conditions (Akimtsev et al., 1962; Agafonov, 1994).

The content of all heavy-metal groups and fractions in soils with different carbonate contents increases under contamination with metals (Table 5.4).

The proportions of metal compounds change in this case (Table 5.5).

Table 5.1. Group composition of Cu, Pb, and Zn in clean chernozem with different carbonate contents, mg/kg

The content of carbonates in soil, %	Weakly bound compounds		Specifically sorbed		Strongly bound compounds			Sum of fractions
	Exchangeable AAB/MgCl$_2$	Complex	on carbonates	on Fe and Mn (hydr)oxides	With organic matter	With Fe and Mn (hydr)oxides	With silicates	
Cu								
0,5	0,3±0,1/ 0,3±0,1	0,2±0,06	1,7±0,4	0,2±0,05	4,2±1,0	0,9±0,1	36,9±5,2	44,4±5,9
2,5	0,4±0,2/ 0,1±0,1	0,1±0,1	2,3±0,5	---	3,8±1,1	3,8±0,4	39,1±3,3	50,0±7,9
5	0,6±0,1/ ---	---	2,7±1,2	0,4±0,1	4,6±0,6	5,0±0,6	44,9±4,0	57,6±5,5
Pb								
0,5	0,6±0,1/ 0,4±0,2	0,3±0,1	1,6±0,5	0,7±0,2	6,5±1,1	1,8±0,08	14,3±2,1	25,6±4,7
2,5	0,4±0,1/ ---	0,3±0,06	1,7±0,7	1,3±0,4	5,6±1,0	4,5±0,4	14,0±1,8	27,4±3,9
5	0,6±0,1/ 0,2±0,1	-	2,6±1,0	0,3±0,1	5,1±1,5	5,8±0,4	16,3±2,7	30,3±4,7
Zn								
0,5	0,4±0,1/ 0,3±0,04	0,3±0,04	6,3±1,3	0,3±0,1	1,0±0,3	6,2±1,6	55,9±4,4	70,3±7,0
2,5	0,2±0,06/ ---	---	7,0±2,0	0,5±0,06	1,0±0,4	10,6±1,8	60,9±3,4	80,0±5,6
5	0,2±0,1/ ---	---	7,5±2,0	---	0,7±0,3	14,9±1,4	65,7±3,0	88,8±5,0

Table 5.2. Effect of carbonates on the group composition of Cu, Pb, and и Zn in clean chernozem (% of the sum of fractions)

The content of carbonates in soil, %	Weakly bound compounds		Specifically sorbed		Strongly bound compounds		
	Exchangeable, MgCl₂	Complex	on carbonates	on Fe and Mn (hydr)oxides	With organic matter	With Fe and Mn (hydr)oxides	With silicates
Cu							
0,5	0,7	0,5	3,8	0,5	9,5	2,0	83,1
2,5	0,2	0,2	4,6	0,0	7,6	7,6	79,8
5	0	0	4,7	0,7	9,7	12,2	72,7
Pb							
0,5	1,6	1,2	6,3	2,7	25,4	7,0	55,9
2,5	0	1,1	6,2	4,7	20,4	16,4	51,1
5	0,7	0	8,6	1,0	16,8	19,1	53,8
Zn							
0,5	0,4	0,4	9,0	0,4	1,4	8,8	79,5
2,5	0	0	8,8	0,6	1,3	13,3	76,1
5	0	0	8,4	0	1,9	19,0	70,6

Table 5.3. Relative contents of Cu, Pb, and Zn forms in groups of weakly and strongly bound compounds in the clean chernozem, %

The content of carbonates in soil, %	Weakly bound compounds				Strongly bound compounds			WB/SB
	Exchangeable, MgCl$_2$	Complex	Specifically sorbed		With organic matter	With Fe and Mn (hydr)oxides	With silicates	
			on carbonates	on Fe and Mn (hydr)oxides				
Cu								
0,5	13	8	71	8	10	2	88	5/95
2,5	4	4	92	-	8	8	84	5/95
5	-	-	87	13	9	9	83	5/95
Pb								
0,5	13	10	53	23	29	8	63	12/88
2,5	-	9	52	39	23	19	58	12/88
5	6	-	84	10	19	21	60	10/90
Zn								
0,5	4	4	88	4	2	10	89	10/90
2,5	-	-	93	7	1	15	84	9/91
5	-	-	100	-	1	18	81	8/92

Table 5.4. Group composition of Cu, Pb, and Zn in contaminated chernozem with different carbonate contents, mg/kg

The content of carbonates in soil, %	Weakly bound compounds			Specifically sorbed		Strongly bound compounds			Sum of fractions
	Exchangeable, AAB/MgCl₂	Complex	On carbonates	on Fe and Mn (hydr)oxides	With organic matter	With Fe and Mn (hydr) oxides	With silicates	
Cu								
0,5	14,0±3,9/ 7,9±2,0	44,9±4,2	36,3±4,9	18,6±5,4	106,5±10,2	76,1±7,8	69,2±4,8	359,7±18,8
2,5	9,6±1,4/ 3,0±2,4	30,7±3,5	41,5±6,8	10,0±4,3	95,0±8,0	87,1±5,9	75,8±3,9	343,1±11,4
5	6,0±1,6/ 1,9±1,3	15,9±4,0	45,0±6,0	5,6±5,5	102,7±6,8	102,1±5,8	78,1±6,6	351,3±18,8
Pb								
0,5	12,4±1,5/ 10,2±2,5	46,0±5,0	32,8±4,0	12,0±3,5	114,2±12,7	74,9±7,9	36,9±4,3	327,0±19,0
2,5	8,0±1,9/ 4,2±1,7	25,0±6,3	39,0±5,6	2,0±5,1	114,2±15,0	108,8±6,9	47,7±3,7	340,9±20,2
5	4,2±2,5/ 1,5±0,5	10,5±6,0	41,9±4,0	3,0±5,7	135,3±11,9	115,2±5,8	43,7±6,0	351,1±14,0
Zn								
0,5	22,5±3,7/ 15,2±3,0	20,7±3,6	50,7±5,2	40,0±5,6	3,9±1,1	104,5±11,3	84,0±5,7	319,0±18,0
2,5	10,5±3,7/ 6,0±3,2	11,7±3,0	57,9±4,9	10,0±6,4	10,1±3,3	120,0±7,4	95,2±5,0	310,9±12,4
5	6,8±1,7/ 2,0±0,6	2,0±0,6	58,8±6,9	5,8±6,1	9,9±3,0	136,6±10,0	90,7±4,6	305,8±15,7

Table 5.5. Group composition of Cu, Pb, and Zn in contaminated chernozem with different carbonate contents, % of the sum of fractions

The content of carbonates in soil, %	Weakly bound compounds		Specifically sorbed		Strongly bound compounds		
	Exchangeable, MgCl$_2$	Complex	on carbonates	on Fe and Mn (hydr)oxides	With organic matter	With Fe Fe and Mn (hydr)oxides	With silicates
Cu							
0,5	2,2	12,5	10,1	5,2	29,6	21,2	19,2
2,5	0,9	8,9	12,1	2,9	27,7	25,4	22,1
5	0,5	5,9	12,8	1,6	29,2	28,2	21,7
Pb							
0,5	3,1	14,1	10,0	3,7	34,9	22,9	11,3
2,5	1,2	7,3	11,4	0,6	33,5	31,9	14,0
5	0,4	3,0	11,9	0,9	38,5	32,8	12,4
Zn							
0,5	4,8	6,5	15,9	12,5	1,2	32,8	26,3
2,5	1,9	3,8	18,6	3,2	3,2	38,6	30,6
5	0,7	0,8	19,9	2,0	3,2	42,8	30,6

Table 5.6. Relative contents of Cu, Pb and Zn forms in groups of weakly and strongly bound compounds in the contaminated chernozem

The content of carbonates in soil, %	Weakly bound compounds			Strongly bound compounds				WB/SB
	Exchangeable, MgCl$_2$	Complex	Specifically sorbed		With organic matter	With Fe and Mn (hydr) oxides	With silicates	
			On carbonates	on Fe and Mn (hydr)oxides				
Cu								
0,5	7	42	34	17	42	30	27	30/70
2,5	4	36	49	12	37	34	29	25/75
5	3	23	66	8	36	36	28	19/81
Pb								
0,5	10	46	32	12	51	33	16	31/69
2,5	6	36	56	3	42	40	18	21/79
5	3	18	74	5	46	39	15	16/84
Zn								
0,5	12	16	40	32	2	54	44	40/60
2,5	7	14	68	12	4	53	42	28/72
5	3	3	86	8	4	55	41	22/78

The presence of carbonates in the contaminated soils even more hampers the increase in metal mobility compared to their clean analogues (Tables 5.3, 5.6). The mobility of Cu, Pb, and Zn decreases by almost two times, when the content of carbonates increases from 0.5 to 5% $CaCO_3$. A similar increase is observed for the heavy metal compounds specifically sorbed by carbonates, and the content of other mobile fractions (exchangeable, complex, and specifically sorbed by Fe and Mn (hydr)oxides) decreases.

The share of Cu, Pb, and Zn compounds specifically sorbed by carbonates in contaminated soils is lower than that in the analogous treatments of clean soils. This can be related not only to the larger proportions of other fractions in the group composition of metal contents, but also to the formation of heavy metal carbonates.

It was shown in Chapter 3 that the accumulation of elements in the fractions bound to carbonates and Fe–Mn oxides directly depended on the concentration of heavy metals in the soil. At the same time, the share of a carbonate-bound metal in the group composition decreased with increasing its content. This could be related to the formation of difficultly soluble heavy-metal carbonates. A 1 M $NaCH_3COO$ solution with pH 5.0 cannot completely dissolve metal carbonates. After the treatment of metal-contaminated soils with this reagent, a part of carbonates (5 to 15%) remains in the soil, in distinction from their complete dissolution in the clean soil.

The changes in the group composition of metals under the introduction of additional amounts of carbonates occur due to the increase in the sorption capacity of soil according to several possible mechanisms:

1) An appreciable part of heavy metal (HM) ions is sorbed by carbonates through specific sorption (chemisorption):

$$HM^{2+} + CaCO_3 = HMCO_3 \text{ (ads)} + Ca^{2+}$$

The binding of heavy metal ions via chemisedimentation in the form of low-mobile carbonates plays an important role in calcareous soils (Glazovskaya, 1994). In ordinary chernozem, this process is favored by the presence of micellar $CaCO_3$ and the nonpercolative water regime. The precipitation of heavy metals on the surface of carbonates blocks the active centers of crystals and inhibits the dissolution of $CaCO_3$.

2) In the presence of high concentrations of heavy metals, when the entire surface of carbonate is covered by chemisorbed metal, the metal begins to precipitate as a separate solid phase of metal carbonate:

$$HM^{2+} + H_2CO_3 = HMCO_3 \text{ (solid)} + 2H^+$$

3) An increase in the content of carbonates to 5% increases the pH by 0.5, from 7.4 to 7.9. The negative charge of the soil increases in this case, because the ionization of functional groups in humus acids is enhanced, the negative charge of clay minerals increases, and the positive charge of Fe and Mn (hydr)oxides decreases (Gorbatov, Zyrin, 1988). Fe and Al hydroxides are ampholites and manifest acid properties with increasing pH, which favors the sorption of metals (Motuzova, Popova, 1988). Pot experiments showed (Table 3.18) that the presence of carbonates in soils increases the sorption activity of nonsilicate Fe and Mn compounds and their participation in the strong fixation of metals.

(4) Another mechanism is related to the formation of hydroxo complexes of heavy metals (HMOH$^+$) with a lower charge at increased pH, which increases their sorption by the soil from solution (Minkina et al., 2005).

Thus, the presence of carbonates in ordinary chernozems and their interaction with heavy metal ions are of great importance for the development of ecologically valuable properties of regional soils. The diversity of carbonate forms in soils and their polyfunctionality ensure the formation of both weakly and strongly bound metal compounds. In chernozems contaminated with Cu, Zn, and Pb, carbonates hamper the increase in metal mobility.

5.3. EFFECT OF AMELIORANTS ON THE GROUP COMPOSITION OF ZINC AND LEAD COMPOUNDS IN CONTAMINATED CHERNOZEM

The effect of different ameliorants on ordinary chernozem contaminated with heavy metals was studied in a long-term field experiment in the Rostovskii state crop testing site, which was described in Section 3.1.2.1. Ameliorants were applied 2 months after the contamination of soils with Zn and Pb acetates. Chalk (2.5 and 5 mg/m^2), glauconite (2 kg/m^2), and semidecomposed cattle manure (5 kg/m^2) were used as ameliorants, as well as their combinations according to the following experimental design:

(1) without ameliorants;
(2) metal (Me);
(3) Me + glauconite;
(4) Me + manure;
(5) Me + glauconite + manure;
(6) Me + chalk, 2.5 kg/m^2;
(7) Me + chalk, 5 kg/m^2;
(8) Me + chalk, 2.5 kg/m^2 + manure;
(9) Me + chalk, 5 kg/m^2 + manure.

Experiments were conducted in triplicate. Soil was artificially contaminated in 2000; three months later, the above ameliorants were applied. The study of their aftereffect was started one year later and continued up to 2004. Samples were taken at a depth of 0–20 cm.

The chalk used in the experiment contained the following metal contents (mg/kg): Zn, 10.3; Ni, 2.56; Mn, 17.3; Cr, 17.4. As, Pb, Cu, and Fe were not found in the samples (Table 5.7).

Table 5.7. Heavy metal content in the chalk, mg/kg

As	Pb	Zn	Cи	Ni	Fe	Mn	Cr
-	-	10,3	-	2,56	-	17,3	17,4

Based on the results of pot experiment on soils with different contents of carbonates, the following application rates of chalk were used in the field experiment on soil remediation: 2.5 and 5 kg/m^2.

Glauconite was used as a natural sorbent. Glauconite (from the Greek glaucos – bluish-green) is a complex potassium-containing aluminosilicate, mineral of hydromica group, layered silicate subclass, with the formula K[(Fe^{3+}, Al)$_{1,33}$ (Fe^{2+}, Mg)$_{0,67}$]$_2$ Si$_{3,67}$ Al$_{0,33}$ O$_{10}$(OH)$_2^-$.

Rock mineralogy: fraction >0.01 mm, 75% (including glauconite, 42%; quartz, 32%); fraction <0.01 mm, 25% (including glauconite, 24%; quartz, 1%). Montmorillonite, calcite, sponge spicules, and siliceous organism residues were not found in the samples. The total content of glauconite in the rock was 66%.

Bulk chemical composition of rocks: SiO_2, 69.5%; Al_2O_3, 4.9%; Fe_2O_3, 13.8%; FeO, 0.20%; CaO, 1.35%; MgO, 1.94%; MnO, 0.016%; K_2O, 3.60%; Na_2O, 0.18%; P_2O_5, 0.41%; TiO_2, 0.30%; SO_3(tot), 0.10%; CO_2, not det., B_2O_3, 0.06%; W_{hygr}, 2.08%; organic matter, 0.23%. The content of heavy metals in glauconite was as follows: Li, 0.0018%; Zn, 0.0034%; Cu, <0.001%; Pb, <0.01%; Ni, 0.004%; Cr, 0.0164%; V, <0.01%; F, 0.16%; Co, <0.001%; Mn, 0,010%.

In the experimental glauconite samples, Pb was not found and the content of Zn was 0.1 mg/kg. With account for the application rate of 2 kg/m^2, 2 mg/kg Zn arrived to the soil with glauconite.

Semidecomposed cattle manure had the following parameters: water, 65.2%; ash, 26.2%; pH, 7.5; N_{tot}, 0.85%; P_2O_{5tot}, 0.87%; K_2O_{tot}, 0.90%.

Total heavy metals in manure: Zn, 60.0 mg/kg; Pb, 12 mg/kg; Cd, 4 mg/kg; Cu, 9 mg/kg. With account for the organic fertilizer rate (5 kg/m^2), 3 mg/kg Zn and 0.6 mg/kg Pb were applied to the soil.

The analysis of the group composition of heavy metal compounds in the reclaimed soils revealed the mechanism of ameliorant action on the mobility of heavy metals in soils.

The total content of metals in the soils with ameliorants remained almost similar to that in the contaminated soils (Table 5.8). However, the group of weakly bound compounds and, hence, the mobility of metals decreased (Tables 5.8, 5.9) because of the decrease in the absolute content of all mobile Zn and Pb forms: exchangeable, complex, and specifically sorbed ones (Table 5.10). The effects of chalk, manure, and glauconite on the mobilization of metal in the soil were different.

Already in the first year after the application of chalk, in distinction from other ameliorants, the content of exchangeable Zn and Pb forms in the soil decreased to the MCL level and lower. The share of weakly bound compounds (Table 5.9) and the value of Km (Table 5.10) decreased to a higher extent than at the application of glauconite or manure.

The main difference between the effects of 2.5 and 5% chalk on the mobility of metals was that the share of specifically sorbed forms in the group composition of Zn and Pb increased with increasing application rate of chalk. The content of exchangeable compounds decreased respectively.

The effect of glauconite and manure applied separately on the fixation of Zn and Pb was lower, which could be related to the formation of their metal compounds with lower binding strength and insufficient interaction time.

Manure organic matter favored the relatively rapid formation of unstable zinc complexes, which were decomposed already to the second year after application, and the released metal passed into more stable organomineral forms.

For lead, the formation rates of metal complexes with manure organic substances were lower; however, their amount significantly increased for three years (Tables 5.9, 5.11). The content of specifically sorbed lead compounds increased because of the reinforcement of their bond with soil components and transition into strongly bound compounds.

It should be noted that the change in mobility of heavy metals in the soil at the application of manure was related not only to the formation of metal-organic compounds, but also to the degree of organic matter decomposition. The mineralization of semidecomposed manure in the soil could be accompanied by the formation of water-soluble low-molecular-weight organic complexes, which increased the migratory capacity of metals. As organic matter was decomposed, the immobilizing effect became more manifested due to the formation of more strongly bound compounds of heavy metals with organic matter. A similar phenomenon was observed at the application of fresh manure and undecomposed and semidecomposed straw to the soil (Evdokimova, 1985; Sizov et al., 1990; Chernykh et al., 1995).

According to the efficiency of the separate application of ameliorants in the fixation of Zn and Pb in the soil, they form the series: chalk > glauconite ≈ manure. Thus, chalk is the best sorbent for Pb, as well as for Zn.

The following possible mechanisms of its effect on metals in contaminated soils are supposed:

1. chemisorption of metals on the surface of $CaCO_3$ particles;
2. formation of separate solid phases (precipitates of heavy metal carbonates and hydroxocarbonates);
3. increase in the sorption activity of Fe, Al, and Mn oxides in the presence of carbonates; and
4. formation of heavy metal hydroxocomplexes at increasing pH, which can increase their sorption by the soil.

The action of the above mechanisms was described in detail in the previous Section. The dissolution of chalk in the field and model experiments mainly occurred in the first year after application; insignificant changes were observed in the following years (Minkina et al., 2007), which pointed to the possible formation of chemically sorbed metal carbonates and the further stabilization of the calcium–carbonate system.

Table 5.8. The total content (numerator) and the ratio between the weakly and strongly bound Zn and Pb compounds (denominator) during 3 years after the application of ameliorants

Experimental treatments	Zn			Pb		
	1 year	2 year	3 year	1 year	2 year	3 year
Without metal addition	68 12/88	65 12/88	67 12/88	24 15/85	24 15/85	28 12/88
Metal (Me)	356 32/68	349 35/65	352 36/64	110 38/62	101 45/55	100 42/58
Me + glauconite	346 14/86	358 13/87	344 7/93	106 32/68	111 31/69	108 16/84
Me + manure	360 17/83	360 12/88	353 10/90	110 26/74	102 27/73	106 24/76
Me + glauconite + manure	363 15/85	365 15/85	352 13/87	109 22/78	101 25/75	111 17/83
Me + chalk, 2.5 kg/m^2	350 13/87	341 14/86	345 10/90	111 18/82	105 31/69	101 12/88
Me + chalk, 5 kg/m^2	347 16/84	342 16/84	344 9/91	108 17/83	112 23/77	103 17/83
Me + chalk, 2.5 kg/m^2 + manure	361 11/89	346 12/88	353 5/95	105 19/81	112 18/82	110 16/84
Me + chalk, 5 kg/m^2 + manure	359 11/89	350 11/89	353 6/94	114 13/87	106 16/84	104 11/89
LSD$_{0,95}$ for numerator	12,3	9,0	8,3	7,5	10,3	9,5

Table 5.9. Weakly bound Zn and Pb compounds in ordinary chernozem during 3 years after the application of ameliorants, mg/kg

Experimental treatments	Compounds									
	Exchangeable			Complex			Specifically sorbed		
	1 year	2 year	3 year	1 year	2 year	3 year	1 year	2 year	3 year
Zn									
Without metal addition	0,6	0,6	0,6	0,4	0,5	0,4	6,5	6,8	6,9
Metal (Me)	33,0	27,6	26,1	27,9	23,7	24,4	52,3	69,8	76,6
Me + glauconite	27,8	16,2	6,9	8,1	6,4	1,36	14,4	24,1	16,4
Me + manure	25,1	18,1	8,4	14,5	5,7	7,6	20,3	17,7	19,1
Me + glauconite + manure	25,4	12,6	3,6	10,6	5,6	13,2	19,9	36,2	29,5
Me + chalk, 2,5 kg/m^2	21,6	9,3	3,9	7,8	7,2	3,0	17,2	30,9	29,3
Me + chalk, 5 kg/m^2	18,0	4,5	1,0	8,8	14,0	4,88	29,9	37,8	25,0
Me + chalk, 2,5 kg/m^2 + manure	15,4	6,8	4,7	8,2	6,15	0,5	15,3	27,4	12,1
Me + chalk, 5 kg/m^2 + manure	10,2	4,0	1,0	9,8	6,0	4,05	20,0	26,5	13,2
LSD$_{0,95}$	6,4	8,0	2,3	1,4	2,01	2,2	10,3	9,9	5,1
Pb									
Without metal addition	0,8	0,9	1,0	0,3	0,3	0,1	2,4	2,5	2,2
Metal (Me)	12,8	10,8	8,7	6,0	12,8	14,7	22,9	21,7	18,3
Me + glauconite	8,4	6,1	3,0	4,9	8,7	4,8	20,9	19,4	9,5
Me + manure	8,0	6,6	5,3	3,6	4,0	7,5	17,1	16,9	12,4
Me + glauconite + manure	7,2	5,3	1,5	1,8	6,8	6,6	14,9	13,1	10,5
Me + chalk, 2,5 kg/m^2	6,7	3,2	2,7	4,5	9,5	2,6	9,3	20,0	6,8
Me + chalk, 5 kg/m^2	4,6	3,0	0,9	4,2	5,4	1,8	9,5	17,7	14,5
Me + chalk, 2,5 kg/m^2 + manure	5,7	2,2	1,5	1,5	7,3	5,1	12,9	11,1	10,5
Me + chalk, 5 kg/m^2 + manure	4,5	2,0	1,0	0,2	4,9	3,4	9,8	10,0	7,1
LSD$_{0,95}$	1,4	4,0	1,1	1,3	2,3	1,6	9,9	10,5	4,0

Table 5.10. Mobility (Km) of Zn and Pb in the soil during 3 months after the application of ameliorants

Experimental treatments	Km Zn 1 year	2 year	3 year	Km Pb 1 year	2 year	3 year
Without metal addition	0,1	0,1	0,1	0,2	0,2	0,1
Metal (Me)	0,5	0,5	0,6	0,6	0,8	0,7
Me + glauconite	0,2	0,2	0,1	0,5	0,4	0,2
Me + manure	0,2	0,1	0,1	0,4	0,4	0,3
Me + glauconite + manure	0,2	0,2	0,2	0,3	0,3	0,2
Me + chalk, 2.5 kg/m^2	0,2	0,2	0,1	0,2	0,5	0,1
Me + chalk, 5 kg/m^2	0,2	0,2	0,1	0,2	0,3	0,2
Me + chalk, 2.5 kg/m^2 + manure	0,1	0,1	0,1	0,2	0,2	0,2
Me + chalk, 5 kg/m^2 + manure	0,1	0,1	0,1	0,1	0,2	0,1

Table 5.11. Weakly bound Zn and Pb compounds (numerator) and the ratio of their exchangeable, complex, and specifically sorbed forms (denominator) during 3 years after the application of ameliorants

Experimental treatments	Zn			Pb		
	1 year	2 year	3 year	1 year	2 year	3 year
Without metal addition	8 8/5/87	8 8/6/86	8 8/5/87	4 23/9/68	4 24/8/68	3 30/3/67
Metal (Me)	113 29/25/46	121 23/19/58	127 21/19/60	42 31/14/55	45 24/28/48	42 21/35/44
Me + glauconite	50 55/16/29	47 35/14/51	25 28/6/67	34 25/14/61	34 18/25/57	17 17/28/55
Me + manure	60 42/24/34	42 43/14/43	35 24/22/54	29 28/12/60	28 24/15/61	25 21/30/49
Me + glauconite + manure	56 45/20/35	55 23/11/66	46 8/28/64	24 30/8/62	25 21/27/52	19 8/36/56
Me + chalk, 2.5 kg/m^2	47 46/17/37	47 20/15/65	36 11/8/81	21 33/22/45	33 10/29/61	12 22/22/56
Me + chalk, 5 kg/m^2	57 32/16/52	56 8/25/67	31 3/16/81	18 25/23/52	26 11/21/68	17 5/11/84
Me + chalk, 2.5 kg/m^2 + manure	39 40/20/40	40 17/15/68	18 26/6/68	20 28/8/64	21 11/35/54	17 9/30/61
Me + chalk, 5 kg/m^2 + manure	40 25/25/50	37 11/16/73	18 5/22/73	15 31/1/68	17 12/29/59	12 9/29/62
LSD$_{0.95}$ for numerator	10.3	9.9	5.1	9.9	10.5	4.0

Note: Data in numerator and denominator are given in mg/kg and % of the group of weakly bound compounds, respectively.

The formation of separate solid phases of heavy metal carbonates can play a significant role in the immobilization of pollutants, because the addition of carbonates to the contaminated soil under field experimental conditions creates conditions for the precipitation of low-soluble metal compounds.

Thus, it can be supposed that chemisorption and precipitation can be the main mechanisms of Zn and Pb sorption by carbonates in ordinary chernozem. The increase in the sorption activity of Fe–Mn (hydr)oxides in calcareous soils plays an important role in the fixation of pollutants.

The simultaneous application of manure and chalk enhanced the sorption effect and resulted in the maximum decrease in the content of weakly bound Zn and Pb compounds. Already in the first year of their simultaneous use, the group ratio and, hence, the mobility of metals were similar to the analogous parameters of the initial soil (Tables 5.8, 5.10).

This could be related to the fact that the stabilization of organic manure compounds with the formation of stable organomineral complexes occurred in the presence of chalk. These complexes strongly fixed heavy metals and made them unavailable for extraction by the reagents used. The addition of 10% of humic acids to calcite under model experimental conditions increased the adsorption of Zn compared to the pure $CaCO_3$ (Brummer et al., 1983).

The results of studies showed differences in the behavior of Zn and Pb under the application of manure. When the soil was composted with manure, the share of complex Zn forms and specifically sorbed Pb forms increased and the content of weakly sorbed metal compounds decreased in the first year (Table 5.11). In the following two years, the share of complex forms decreased for Zn and increased for Pb. This could be due to the formation of mobile Zn complexes with organic ligands and more stable Pb complexes.

The efficiency of remediation was significantly higher in the second and third years after the application of ameliorants, which was related to the transformation rates of metal compounds and their more complete interaction with the sorbents.

It is suggested that the effect of metal nature on the transformation of its compounds becomes the most important with time, regardless of the ameliorant.

General and specific features were revealed in the transformation of compounds of two metals under the effect of ameliorants. The general tendencies were as follows:

- the application of all ameliorants decreased the mobility of metals;
- the addition of manure to chalk or glauconite increased the binding strength by soils and decreased the content of their exchangeable compounds;
- the highest ameliorating effect was observed at the simultaneous application of chalk and manure to the contaminated soils;
- in the contaminated soils and those treated with ameliorants, most of the weakly bound metal compounds were specifically sorbed forms representing the nearest reserve of strongly bound metal compounds;
- the transformation of metal compounds occurred in the contaminated soil during three years after the application of ameliorants, which resulted in a decrease in their mobility and in the share of the most migration-capable exchangeable forms and the corresponding increase in the content of strongly bound forms.

The difference in the state of soils polluted with lead and zinc salts contained in ameliorants was that the rates of zinc fixation in soils were higher. Therefore, the share of strongly bound zinc compounds reached and even exceeded the values typical for clear soils (Table 5.8). As for lead, only the simultaneous application of chalk and manure increased the relative content of its strongly bound forms to the level of the initial soil.

Thus, the application of ameliorants significantly decreased the mobility of metals. Their effect depended on the ameliorant and was most significant at the simultaneous application of chalk and manure. This effect was presumably due to the strong binding of metals by carbonates through chemisorption and formation of low-soluble Zn and Pb compounds and to the additional fixation in the form of complexes at the addition of organic material. Therefore, the share of weakly bound metal compounds in the contaminated soils decreased to the level typical for the clean soils or even below (in the case of zinc). The general evolution of the transformation of metal compounds (from less to more strongly bound compounds) accelerated by ameliorants remained for both metals, but the rates of these processes for Zn compounds were higher than for Pb compounds.

CONCLUSIONS

The system of methodological approaches used for determining the group composition and mobility of heavy metal compounds in soils was found to be efficient in the assessment of soil status in the Lower Don region. The results are of ecological significance, because they allow predicting the transformation of metal compounds retained by different soil components and assessing the probability of their secondary mobilization. Further development of the proposed method will contribute to the assessment of heavy metal status in soils.

The fractionation of heavy metal compounds from contaminated and clean soils of the Lower Don region in a series of model laboratory, pot, and field experiments and monitoring observations showed that the group composition of heavy metal compounds is indicative of pedogenesis conditions, anthropogenic load on soils, and soil ecological functions.

The polyfunctionality of soil components (organic substances, carbonates, Fe–Mn (hydr)oxides) is manifested in their capacity for strong and weak fixation of metals.

Soils of the Lower Don region are characterized by the increased total contents of Cu, Zn, and Pb with a predominance (56–83%) of their strongly bound forms in the structure of primary and secondary minerals. Weakly bound heavy-metal compounds predominantly retained by carbonates constitute 5–12%.

The contamination of soils with Cu, Pb, and Zn disturbs natural ratios of metal compounds: the mobility of metals increases (especially under polymetal contamination), and the soil tolerance toward the impact of metals decreases. Organic substances are most active in the weak fixation of copper and lead; Fe–Mn (hydr)oxides are also active in the retention of zinc. The strong fixation of applied copper and lead is ensured by organic substances and Fe–Mn (hydr)oxides, the strong fixation of zinc is mainly due to Fe–Mn (hydr)oxides. The fixation of applied metals in the lattices of silicate minerals is insignificant.

Metals in soils represent a continual series of compounds with a common direction of transformation from less stable to more stable forms. The initial transformation of metal compounds is mainly related to the transition of exchangeable forms into specifically sorbed forms for Zn and from exchangeable forms into complexes with organic substances for Cu and Pb. Zinc ions are less strongly fixed and are weak competitors for adsorption sites in the presence of copper and lead.

No equilibrium in the system of heavy metal compounds is reached in ordinary chernozem during 3–5 years after contamination. Under polymetal contamination, the mobility of metals in the soil is higher than that under monometal contamination. The long-term (more than 40 years) industrial contamination of soils with heavy metals has led to a significant increase in the content of their mobile compounds in comparison with the single application of metals.

With respect to the capacity for strong fixation of Cu, Pb, and Zn, the soils in the Lower Don reaches compose the following sequence: sandy clay meadow-chernozemic soil ≥ clay loamy meadow-chernozemic soil > clay loamy ordinary chernozem > clay loamy chestnut soil > alluvial meadow soil. A higher contribution of exchange processes to the transformation of heavy metals is observed in chestnut soils in comparison with chernozems. The capacity of soils to strongly fix the metals increases with an increase in the content of soil carbonates.

REFERENCES

[1] Abd-Elfattan A., Wada K. Adsorbtion of lead, copper, zinc, cobalt, and cadmium by soils that dipper in cation-exchange material // *J. Soil Sci.* - 1981. - V. 32. - P. 271-283.

[2] Adriano D.C. Trace Elements in Terrestrial Environments. - New York: Springer-Verlag, 2001. – 868 p.

[3] Agafonov E.V. Heavy metals in chernozems of Rostov oblast // Heavy Metals and Radionuclides in Agroecosystems: Collection of Articles. Novocherkassk, 1994. P. 22–26.

[4] Ageev V.N., Val'kov V.F., Cheshev A.S., Tsvylev E.M. Ecological Aspects of Soil fertility in Rostov Oblast // Rostov-on-Don: SKVSh, 1996. - 167 p.

[5] Agrochemical Methods of Soil Study. - Moscow: Nauka, 1975. - 656 p.

[6] Ahumada I., Escudero P., Carrasco A., Castillo G., Sadzawka A. Sequential extraction of some heavy metals from Chilean soils amended with sewage sludge // *Proceedings of the 7th International Conference on the Biogeochemistry of Trace Elements,* Uppsala, Sweden, June 15-19, 2003. - V. 1. - P. 124-125.

[7] Akimtsev V.V., Boldyreva A.V., Golubev S.N., Kudryavtsev M.I., Rudenskaya K.V., Sadimenko P.A., Sobornikova I.G. Microelements in soil of Rostov oblast // *Proceedings of the 3rd Interuniversity Workshop "Microelements and Natural Radioactivity of Soils",* – Rostov-on-Don: RGU, 1962. - P. 38-41.

[8] Aleksandrova L.N. Soil Organic Matter and Its Transformation. - Leningrad: *Nauka,* 1980. - 287 p.

[9] Aleksandrova L.N., Dorfman E.M., Yurlova O.V. Organomineral derivatives of humic substances in the soil // *Soil Humus Substances.* – Leningrad–Pushkin, 1970. - V. 142. - P. 157-197.
[10] Aleksandrova L.N., Naidenova O.A. Composition and nature of soil humus substances // *Soil Humus Substances.* – Leningrad–Pushkin, 1970. - V. 142. - P. 83-140.
[11] Alekseev Yu.V. Vyalushkina N.I. Effect of calcium and magnesium on the input of cadmium and nickel from the soil into vetch and barley plants // *Agrokhimiya.* - 2002. – No. 1. - P. 82-84.
[12] Amit K. G., Chen Z.S. Role of EDTA and NTA in the phytoextraction of Cd, Pb, and Zn by *Chenopodium album* L. from artificially contaminated soil // *Proceedings of the 14th International Conference on Heavy Metals in the Environment.* - Taipei, Taiwan, 2008. - P. 591-594.
[13] Andersen M.K., Raulund-Rasmussen K., Strobell B. W., Hansen H.C.B. The effects of tree species and site on the solubility of Cd, Cu, Ni, Pb, and Zn in soils // *Water Air Soil Pollut.* – 2004. – V. 154. – P. 357–370.
[14] Andersson A. On the determination of ecologically significant fractions of some heavy metals in soils // *Swed. J. Agr. Res.* - 1976. - V. 6. No. 1. - P. 197-199.
[15] Andersson A. The distribution of heavy metals in soils and soil material as influenced by ionic radius // *Swed. J. Agric. Res.* - 1977. - No. 7. - P. 141-147.
[16] Anthropogenic Effect of Emissions from the Novocherkassk Power Station on the City and Station Environments: Research Report (NGTsEIiM). - Novocherkassk, 1995. – 38 p.
[17] Antipov-Karataev I.N., Kader G.M. The nature of ion uptake by clays and soils // *Kolloid. Zh.* - 1947. - No. 9. - P. 81-124.
[18] Arinushkina E.V. Handbook on the Chemical Analysis of Soils. - Moscow: MGU, 1961. - P. 490.
[19] Arinushkina E.V. Handbook on the Chemical Analysis of Soils. - Moscow: MGU, 1970. - 488 p.
[20] Arzhanova E.S., Elpat'evskii P.V., Vertel E.F., Microelements and soluvle organic matter of lysimetric waters // *Pochvovedenie.* – 1981. - No. 11. - P. 50-60.
[21] Baidina N.L. Inactivation of heavy metals by humus and zeolites in the technogenically contaminated soil // *Pochvovedenie.* - 1994. - No. 9. - P. 121-125.
[22] Barbu C., Sand C., Oprean L. Use of natural zeolites to remediate soils polluted with heavy metals // *Proceedings of the 7th international*

Conference on the Biogeochemistry of Trace Elements, Uppsala, Sweden, June 15-19, 2003. - V. 1. - P. 250-251.

[23] Baron U. Gemeinsame Extraction und chemische Bestimmung des leicht-loslichen Anteils der Mikronohrstoffe Bor, Eisen, Kobalt, Kupfer, Mangan, Molibden, Zink im Boden. – Landwirtshaftliche Forsuchung, 1955. - Bd. 82. - N 7. – H. 2.

[24] Bartashevsky Yu.A., Gaydarov O.L., Gordienko S.A. Free radical damping investigation of humic acid with different Mn and Cu content ESR method // *Humus Planta*. - 1971. - No. 5. - P. 339-342.

[25] Basta N.T., Armstrong F.P., Hanke E.M. Effect of chemical remediation of contaminated soil on arsenic mobility and gastrointestinal availability // *Proceedings of the 6th International Conference ICOBTE-2001 "Biogeochemistry of Trace Elements"*. – Canada, 2001. – P. 40.

[26] Belousov V.S. Zeolite-containing rocks of Krasnodar krai // *Agrokhimiya*. - 2006. - No. 4. - P. 78-84.

[27] Berti W.R. and Jacobs L.W. Heavy metals in the environment chemistry and phytotoxicity of soil trace elements from repeated sewage sludge applications // *Environ. Qual*. - 1996. - Vol. 25. - P. 1025-1032.

[28] Bezuglova O.S. Humus Status of Soils in the Southern Russia. – Rostov-on-Don: *SKNTs VSh*, 2001. - 228 p.

[29] Bezuglova O.S., Ignatenko E.L., Morozov I.V. Effect of brown coal on the mobility of copper and lead in ordinary chernozem // *Pochvovedenie*. – 1996. – No. 9. – P. 1103-1106.

[30] Bibak A. Cobalt, copper, and manganese adsorption by aluminum and iron oxides and humic acids // *Commun. Soil Sci. Plant Anal*. - 1994. - V. 25, Nos. 19-20. - P. 3229-3239.

[31] Bilali L.E., Rasmussen P.E., Fortin D. Humic substances in sediments of three remote lakes in Ontario and Nova Scotia, Canada: transformations with depth and interactions with trace metals. // *Proceedings of the 6th International Conference ICOBTE-2001 "Biogeochemistry of Trace Elements"*. – Canada, 2001. - P. 67.

[32] Bingham F.T., Page A.L. Retention of Cu and Zn by montmorillonite // *Soil. Sci. Soc. Am. Proc*. - 1964. - V. 28, No. 3. - P. 351-354.

[33] Brown G.E. Jr., Foster A.L., Ostergren J.D. Mineral surfaces and bioavailability of heavy metals: A molecular-scale perspective // *Proc. Natl. Acad. Sci*. – 1999. – V. 96. – P. 3388–3395.

[34] Brown S., Christiansen B., McGrath S.P., Lombi E., McLaughlin M., Vangronsveld J. Use of soil amendments to restore a vegetative cover on zinc and lead mine tailings // *Proceedings of the 6th International*

Conference ICOBTE-2001 "Biogeochemistry of Trace Elements". – Canada, 2001. – P. 43.
[35] Brummer G., Tiller K.G., Herms U., Clayton P. Adsorption–desorption and/or precipitation–dissolution processes of zinc in soils // *Geoderma.* - 1983. - V. 31, No. 4. - P. 337-354.
[36] Camerlinc R., Kiekens L. Speculation of heavy metals in soils based on change separation // *Plant Soil.* - 1982. - V. 68, No. 3. - P. 331-339.
[37] Chaignon V., Sanchez-Neira I., Jaillard B., Hinsinger P. Relation between copper phytoavailability and chemical properties of 24 calcareous soils from a vine-growing area // *Proceedings of the 6th International Conference ICOBTE-2001 "Biogeochemistry of Trace Elements".* – Canada, 2001. - P. 171.
[38] Chao T.T. Selection dissolution of manganese oxides from soils and sediments with acidified hydroxylamine hydrochloride // *Soil Sci. Soc. Am. Proc.* - 1972. - No. 36. - P. 764-768.
[39] Chemistry of Heavy Metals, Arsenic, and Molybdenum in Soils / Ed. by Zyrin N.G., Sadovnikova L.K. - Moscow: MGU, 1985. – 208 p.
[40] Chen Z., Looi K., Liu J. Chemical remediation methods influence on the uptake of cadmium and lead by vegetables in contaminated soils // *Proceedings of the 5th International Conference of Biogeochemistry of Trace Elements.* – Vienna, Austria, 1999. – P. 1012-1013.
[41] Chernykh N.A., Ovcharenko M.M., Popovicheva L.L., Chernykh I.N. Methods of decreasing the phytotoxicity of heavy metals // *Agrokhimiya.* - 1995. - No. 9. - P. 101-107.
[42] Chesire M.V., Berrow M.L., Goodman B.A., Mundie C.M. Metal distribution and nature of some Cu, Mn, and V complexes in humic and fulvic fractions of soil organic matter // *Geochim. Cosmochim. Acta.* - 1977. - V. 53. - P. 377-382.
[43] Chirikov F.V. Agrochemistry of Potassium and Phosphorus. - Moscow, 1956. - 464 p.
[44] Chlopecka A. Forms of Cd, Cu, Pb, and Zn in soil and their uptake by cereal crops when applied jointly as carbonates //*Water Air Soil Pollut.* - 1996. - No. 25. - P. 69-79.
[45] Choi H. J., Ro H. M., Yun S.I. and Lee M. J. Chemical speciation of Cu and Zn in mine soil as affected by organic chelating agents with different C/N ratio // *Proceedings of the 14th International Conference on Heavy Metals in the Environment*, Taipei, Taiwan, 2008. - P. 515-517.
[46] Chudzhiyan Kh., Karveta R., Fatsek Z. Heavy metals in soils and plants // *Ecological cooperation.* - Bratislava, 1988. - N. 1. - P. 5-24.

[47] Crimme H. Fractiometre exstraction von Rupfer aus Boden // *Z. Pflanzenernaehr.* Bodenk–d. - Bd. 113. - 1967. - H. 43.
[48] Dabakhov M.V. Solov'ev G.A., Egorova V.S. Effect of agrochemicals on the mobility of Pb and Cd in light gray forest soil and their input into plants // *Agrokhimiya*. - 1998. - No. 3. - P. 54-59.
[49] Dahn R., Scheidegger A.M., Manceau A., Schlegel M., Baeyens B., Bradbury H., Morales M. Neoformation of Ni phyllosilicate upon Ni uptake on montmorillonite. A kinetic study by powder and polarized EXAFS // *Geochim. Cosmochim. Acta*. - 2002. - V. 66. – P. 2335-2347.
[50] Davis J.A., Fuller C.C., Cook A.D. A model for trace metal sorption processes on the calcite surface: Adsorption of Cd^{2+} and subsequent solid solution formation // *Geochim. Cosmochim. Acta.* – 1987. – No. 51. – P. 1477–1490.
[51] Demin V.V. Role of humic acids in the irreversible sorption of heavy metals in the soil // *Izv. Timiryazevsk. S--kh. Akad.* – 1994 - No. 2. - P. 79-86.
[52] Determination of the Toxicity of Metal-Contaminated Soils and Some Problems of Its Assessment / Ed. by Kryuchkov V.V. - Apatity: AN SSSR, 1985. - 52 p.
[53] Dhillon S.K., Dhillon K.S. Zinc adsorption by alkaline soils // *J. Ind. Soc. Soil Sci.* - V. 32, No. 2. - P. 250-252.
[54] Distanov U.G., Konyukhova T.P. Natural sorbents and the environmental control // *Khim. Sel'sk. Khoz.* - 1990. - No. 9. - P. 34-39.
[55] Dobrovol'skii V.V. Biospheric cycles of heavy metals and the regulatory role of soil // *Pochvovedenie*. - 1997. - No. 4. - P. 431-441
[56] Donisa C., Mocanu R., Steinnes E. Distribution of some major and minor elements between fulvic and humic acid fractions in natural soils // *Geoderma*. – 2003. – V. 111. – P. 75–84.
[57] Donisa C., Steinnes E., Mocanu R. Distribution of some major and minor elements between fulvic and humic acid fractions in forest soils. // *Proceedings of the 6th International Conference ICOBTE-2001 "Biogeochemistry of Trace Elements".* – Canada, 2001. - P. 62.
[58] Dospekhov B.A. Procedure of Field Experiments. - Moscow: *Kolos*, 1968. - 336 p.
[59] Dumat C., Toinen S., Mariotti A., Morin G., Benedetti M. A study of organic matter turn-over in soils contaminated by heavy metals // *Proceeding of the 10th International Meeting of the International Humic Substances Society (IHSS 10)*. - Toulouse, France, July 2-28, 2000. – P. 495-497.

[60] Ecological Atlas of Rostov Oblast / Ed. by Zakrutkin V.E. – Rostov-on-Don: SKNTs VSh, 2000. - 120 p.
[61] Ecological certificate of Novocherkassk: Report of Lange-Scale Geochemical and Radiometric Studies of Ecological Situation. - Novocherkassk, 1995. - 178 p.
[62] Ecology of Novocherkassk: Problems and Methods of Solution / Ed. by Belousova N.V. - Rostov-on-Don: SKNTs VSh, 2001. - 393 p.
[63] Einax J.W., Nischwitz V. Sampling and sequential extraction under inert conditions for the assessment of actual heavy metal mobility in sediments // *Proceedings of the 6th International Conference ICOBTE-2001 "Biogeochemistry of Trace Elements"*. – Canada, 2001. - P. 56.
[64] El'kina G.Ya. Behavior of copper in the soil–plant system in the Northeastern Europe // *Agrokhimiya*. – 2008. – No. 6. – P. 72-79.
[65] El'kina G.Ya. Tabalenkova G.N.m Kurenkova R.B. Effect of heavy metals on the yield and physiological and biochemical properties of oat. // *Agrokhimiya*. - 2001 - No. 8. - P. 73-78.
[66] Emmirich W.E., Lund L.J., Page A.L., Chang A.C. Solid phase forms of heavy metals in sewage sludge treated soils // *J. Environ. Qual*. - 1982. - V. 11, No. 2. - P. 118-124.
[67] Evdokimova G.A. Determination of Toxicity of metal-Contaminated Soils and Some Methods of Its Decrease. – Moscow: *AN SSSR*, 1985. - 86 p.
[68] Farrah H., Pickering W.H. Influence of clay–solute interactions on aqueous heavy metal ion levels // *Water Air Soil Pollut*. - 1977. - No. 8. - P. 189-197.
[69] Fassbender H.W., Seekamp G. Fractionen und Doslichkeit der Schwermetall Cd, Co, Cr, Ni und Pb im Boden // *Geoderma*. - 1977. - No. 16. - P. 55-69.
[70] Fedorov A.S. Behavior of heavy metals in soils of different genesis // *Current Problems of Soil Pollution: Proceedings of the II International Scientific Conference*. - Moscow: MGU, 2007. – P. 253-256.
[71] Fengxiang X. Han, Arieh S. Biogeochemistry of trace elements in arid environments // *Environ. Pollut*. - 2007. - V. 13. – 366 p.
[72] Filatova E.V. Accumulation forms of heavy metals under landscape-geochemical conditions of the Subarctic Eastern Europe: *Extended Abstract of Candidate's Dissertation in Biology*. - Moscow, 1992 – 16 p.
[73] Ford R.G., Scheinost A.C., Sparks D.L. Frontiers in metal/precipitation mechanisms on soil mineral surfaces // *Adv. Agron*. - 2001. - V. 74. - P. 41-62.

[74] Franklin M.L., Morse J.W. The interaction of manganese (II) with the surface if acclimate in dilute solutions and seawater // *Mar. Chem.* - 1983. - No. 12. - P. 241-254.

[75] Frid A.S. Migration concept of soil nutrient availability to plant roots // *Agrokhimiya.* - 1996. - No. 3. - P. 29-37.

[76] Gapon E.N. Study of exchangeable adsorption: I. Exchange of two ions with similar exchange and sorption capacities // *Zh. Obshch. Khim.* - 1937. - V. 7. - Issue 10. - P. 1468-1473.

[77] Gatenhouse P., Russel D.V., van Moort J.C. Sequential soil analysis in exsploration geochemistry // *J. Geochem. Exsplor.* - 1977. - V. 8, No. 1. - P. 489-499.

[78] Gavrilyuk F.Ya., Val'kov F.Ya., Klimenko G.G. Genesis and appraisal of chernozems in the Lower Don basin and Northern Caucasus // *Scientific Principles of the Rational Use of Chernozems.* – Rostov-on-Don: RGU, 1976. – P. 12-21.

[79] Gedroits K.K. Chemical Analysis of Soils: Selected Works. – Moscow, 1955. - V. 2. - 601 p.

[80] Geebelen W., Nangronsveld J., Clijsters H. Lead immobilization in lead contaminated soils // *Proceedings of the 5th International Conference of Biogeochemistry of Trace Elements.* – Vienna, Austria, 1999. – P. 1016-1017.

[81] Gerasimova V.N. Natural zeolites as adsorbents of oil products // *Chemistry for Sustainable Development.* - 2003. - No. 11. - P. 481-488.

[82] Gibson M.J., Farmer J.G. Multi-step sequential chemical extraction of heavy metals from urban soil // *Environ. Pollut.* - 1986. - Ser. B. - No. 11. - P.117-135.

[83] Ginzburg K.E., Lebedeva L.S. Procedure of determining mineral phosphates in soils // *Agrokhimiya* - 1971. - No. 1. - P. 125-131.

[84] Glasner A., Weiss D. The crystallization of calcite from aqueous solutions and the role of zinc and magnesium ions: I. Precipitation of calcite in the presence of Zn^{2+} ions // *J. Inorg. Nucl. Chem.* – 1980. - No. 42. – P. 655–663.

[85] Glazkova E.A., Strel'nikova E.B., Ivanov V.G. Use of natural zeolites from the Khonguruu field (Japan) for the purification of oil-containing waste water // *Chemistry for Sustainable Development.* - 2003. - No. 11. - P. 849-854

[86] Glazovskaya M.A. Criteria for soil classification by the hazard of lead contamination // *Pochvovedenie.* - 1994. - No. 4. - P. 110-120.

[87] Gleyzes C., Tellier S., Astruc M. Chemical characterization of lead in industrially-contaminated soils // *Proceedings of the 6th International Conference ICOBTE-2001 "Biogeochemistry of Trace Elements"*. – Canada, 2001. - P.57.
[88] Golovatyi R.E. Heavy Metals in Agroecosystems. – Minsk, 2002. – 239 p.
[89] Gorbatov I.S. Transformation and status of zinc, lead, and cadmium in soils: Extended Abstract of Candidate Dissertation in Biology. - Moscow, 1983. - 24 p.
[90] Gorbatov V.S. Stability and transformation of heavy metal (Zn, Pb, Cd) oxides in soils // *Pochvovedenie*. - 1988. - No. 1. - P. 35-42.
[91] Gorbatov V.S., Zyrin N.G. Adsorption of Zn, Pb, and Cd by the soil and the acid–base equilibrium // *Vestn. Mosk. Univ., Ser. 17: Pochvoved.* – 1988 - No. 3. - P. 21-25.
[92] Gordeeva O.N., Belogolova G.A. About the procedure of studying the forms of heavy metals in soils of the Southern Cisbaikalia and the Northeastern China // *Proceeding of the II International Scientific Conference "Current Problems of Soil Contamination"*. - Moscow: MGU, 2007. - V. 2. - P. 191-194.
[93] Gray C.W., McLaren R.G. Soil factors affecting heavy metal solubility in some New Zealand soils // *Water Air Soil Pollut.* – 2006. – V. 175. - P. 3–14.
[94] Grin R.G. Clay Mineralogy. - Moscow, 1959. – 452 p.
[95] Han F.X. Biogeochemistry of Trace Elements in Arid Environments. - *Springer*, 2007 – P. 357.
[96] Hering J.G., Morel F.M. Humic acid complexation of calcium and copper // *Environ. Sci. Technol.* - 1988. - V. 22, No. 10. - P. 1234-1237.
[97] Hildebrand E.E., Blum W.E. Lead fixation by clay minerals // *Naturwissencchaften*. - 1974. - V. 61, No. 4. - P. 169.
[98] Hettiarachichi G.M., Pierzynski G.M. In situ stabilization of soil lead using phosphorus and manganese oxide // *J. Environ. Qual.* 2002. 31. P. 564–572.
[99] Hettiarachichi G.M., Pierzynski G.M., Zwonitzer J. and Lambert M. Phosphorus source and rate effects on cadmium, lead, and zinc bioavailabilies in a metal-contaminated soil // *Extended Abstr., 4th Int. Conf. on the Biogeochem. Trace Elements (ICOBTE)*. 1998. pp. 463–464.
[100] Hodgson J.F., Lindsay W.L., Trieveiler J.F. Micronutrient cation complexing in soil solution: II. Complexing of zinc and copper in

displaced solution from calcareous soil // *Soil Sci. Soc. Am. Proc.* - 1966. - No. 30. – P. 723-726.
[101] Huang P.M., Violante A., Bollag J.-M. and Vityakon P. (Eds.). Soil Abiotic and Biotic Interactions and the Impact on the Ecosystem and Human Welfare. *Science Publishers, Inc.* 2005. 444 p.
[102] Il'in V.B. Assessment of ecological norms for the content of heavy metals in soils // *Agrokhimiya.* - 2000. - No. 9. - P. 74-79.
[103] Il'in V.B. Assessment of heavy metal flux in the soil–crop system // *Agrokhimiya.* - 2006. - No. 3. - P. 52-59.
[104] Il'in V.B. Assessment of soil buffering capacity for heavy metals // *Agrokhimiya.* - 1995. - No. 10. - P. 109 -113.
[105] Il'in V.B. Heavy metals in the Soil–Plant System. – Novosibirsk: *Nauka,* 1991. – 151 p.
[106] Il'in V.B., Stepanova M.D. Protection capacity of the soil–plant system under soil contamination with heavy metals // *Heavy Metals in the Environment.* – Moscow, 1980. – P. 80-85.
[107] Il'in V.B., Syso A.I. Microelements and heavy metals in soils and plants of Novosibirsk oblast. – Novosibirsk: *Ross. Akad. Nauk,* 2001. - 229 p.
[108] Izerskaya L.A., Vorob'eva T.E. Heavy Metal Compounds in Alluvial Soils of the Middle Ob Valley // *Eur. Soil Sci.* - 2000. - No. 1. - P. 49-55.
[109] Kabata-Pendias A., Pendias, H. Trace Elements in Soils and Plants. – Boca Raton: CRC, 1985.
[110] Karnaukhov A.I., Tkachenko V.M., Shestidesyatnaya N.L. Study of copper adsorption by some soils of the Ukrainian SSR // *Pochvovedenie.* - 1989. - No. 11. - P. 118-123.
[111] Karpachevskii L.O., Babanin V.F., Gendler T.S., Opalenko A.A., Kuz'min R.N. Diagnostics of iron minerals by Mössbauer spectroscopy // *Pochvovedenie.* - 1972. - No. 10. - P. 110-120.
[112] Karpova E.A. The Effect of Long-Term Mineral Fertilization on the Status of Iron and Heavy Metals in Soddy-Podzolic Soils // *Eur. Soil Sci.* – 2006. - No. 9. – P. 953-960.
[113] Karpukhin A.I. Complex Compounds of Humic Substances with Heavy Metals // *Eur. Soil Sci.* - 1998. - No.7. - P. 764-771.
[114] Karpukhin A.I., Sychev V.G. Complexes of Soil Organic Substances with Metal Ions. - Moscow: VNIIA, - 2005. – 188 p.
[115] Karpukhin F.I., Shestakov E.I., Chepurina T.A. Migration and transformation of iron in podzolic soils // *Dokl. Timiryazevsk. S--kh. Akad.* - 1980. - Вып. 285. - p.49-54.

[116] Katsnel'son Yu.Ya. Geochemical features of glauconite-containing micronodules in the Rostov oblast and methods of their practical use / *Extended Abstract of Candidate's Dissertation in Geology.* – Rostov-on-Don, 1981. – 25 p.
[117] Kaurichev I.S., Karpukhin F.I., Shestakov E.I. Transformation and migration of manganese in podzolic soils // *Изв. Timiryazevsk.* S--kh. Akad. - 1983. - V. 3. - P. 82-
[118] Kaushansky P., Yariv S. The interactions between calcite particles and aqueous solutions of magnesium, barium, or zinc chloride // *Appl. Geochem.* – 1986. - No. 1. – P. 607–618.
[119] Kazakov L.K. Changes in the structure of areas affected by thermal station // *Vestn. Mosk. Univ., Ser. 5: Geogr.* - 1977. - No. 4. - P. 77-81.
[120] Khardikov A.E., Boiko N.I., Agarkov Yu.V. Zeolites in the Southern Russia // *Lithology and Mineral Resources.* – 1999. - No. 4. - P. 389-399.
[121] Khoroshkin M.N., Khoroshkin B.M. Microelements in Soils and Fodders of Rostov Oblast. - *Persianovka,* 1979. – 39 p.
[122] Kirkham M.B. Cadmium in plants on polluted soils: Effects of soil factors, hyperaccumulation, and amendments // *Geoderma.* – 2006. – V. 137. – P. 19–32.
[123] Kizil'shtein L.Ya., Gofen G.I., Peretyatko A.G., Levchenko R.V. Element impurities in coals, combustion products, plants, soils, and atmosphere in the thermal station region // *Izv. SKNTs VSh.* - 1990. - No. 2. - P. 42-52.
[124] Kliaugiene E., Baltrenas P. Research on heavy metals immobilization in roadside soil by natural zeolites // *Proceedings of the 7th international Conference on the Biogeochemistry of Trace Elements, Uppsala, Sweden,* June 15-19, 2003. - V. 1. - P. 76-78.
[125] Knox A.S., Adriano D.C. Effect of zeolite and apatite on mobility and speciation of metals // *Proceedings of the 5th International Conference of Biogeochemistry of Trace Elements.* – Vienna, Austria, 1999. – P. 990-991.
[126] Knox A.S., Seaman J., Adriano D.C., Pierzynski G. Chemophytostabilization of metals in contaminated soils // *Bioremediation of Contaminated Soil* / Ed. by D.L. Wise et al. – New York: Marcel Dekker, 2000. – P. 811-836.
[127] Kornicker W.A., Morse J.W., Damascenos R.N. The chemistry of Co^{2+} interaction with calcite and aragonite surface // *Chem. Geol.* – 1985. - No. 53. – P. 229–236.

[128] Kosheleva N.E., Kasimov N.S., Samonova O.A. Regression Models for the Behavior of Heavy Metals in Soils of the Smolensk-Moscow Upland // *Eur. Soil Sci.* - 2002. - No. 8. - P. 954-966.
[129] Koval'skii V.V. Geochemically environment, microelements, and response of organisms // Proceedings of the Biogeochemical Laboratory - Moscow: *Nauka*, 1961. - P. 5-23.
[130] Koval'skii V.V., Andrianova G.A. Microelements in Soils of the USSR. - Moscow, 1970. - 179 p.
[131] Krupenikov I.A. Calcareous Chernozems. – Chisinau: Shtiintsa, 1979. – 108 p.
[132] Krupskii N.K., Aleksandrova A.M. Determination of mobile microelements // *Microelements in the Life of Plants, Animals and Humans.* – Kiev, 1964. - P. 34-36.
[133] Kuhn J. Distribution of uranium and selected heave metals in the sediments of the floodplain of the Ploucnice river. - PhD Thesis, Charles University in Prague, 1996. - 278 p.
[134] Kuz'mich M.A., Grafskaya G.A., Khostantseva N.V. Effect of liming on the input of heavy metals to pants // *Agrokhim. Vestn.* - 2000. - No. 5. - P. 28-29.
[135] Kuznetsov N.P., Nikushina T.K., Mazhaiskii Yu.A., Pchelintseva R.A. Heavy metals in soils of Ryazan oblast // *Khim. Sel's. Khoz.* - 1995 - No. 5 - P. 22-25.
[136] Kuznetsov V.A., Shimko G.A. Sequential Extraction Method in Geological Studies. - Minsk: *Nauka i tekhnika,* 1990. – 88 p.
[137] Laboratory Manual on Agricultural Chemistry / Ed. by Mineev V.G. - Moscow: MGU, 1989. - 304 p.
[138] Ladonin D.V. Heavy Metal Compounds in Soils: Problems and Methods of Study // *Eur. Soil Sci.* - 2002. - No. 6. - P. 682-692.
[139] Ladonin D.V. Effect of technogenic contamination on the fractional composition of copper and zinc in soils // *Pochvovedenie.* - 1995. - No. 10. - P. 1299-1305.
[140] Ladonin D.V. Ion Competition in Soils Polluted by Heavy Metals // *Eur. Soil Sci.* - 2000. - No. 10. - P. 1129-1136.
[141] Ladonin D.V. Specific Adsorption of Copper and Zinc by Some Soil Minerals // *Eur. Soil Sci.* - 1997. - No. 12. - P. 1478-1485.
[142] Ladonin D.V., Plyaskina O.V. Fractional composition of copper, zinc, and lead compounds in some soils under polyelement contamination // Vestn. Mosk. Univ., Ser. 17: *Pochvoved.* - 2003. - No. 1. - P. 9-16.

[143] Lamoureux M.M., Warner N., Nizam N., Gordon T., Sullivan E. X-ray absorption fine structure spectroscopy: Shining new lights on metal speciation of environmental solids // *Proceedings of the 6th International Conference ICOBTE-2001 "Biogeochemistry of Trace Elements"*. – Canada, 2001. - P. 120.

[144] Lavrent'eva G.V. Dynamics of heavy metals and macrocations in soil solution at the contamination of leached chernozem with Co, Cu, Zn, and Cd // *Agrokhimiya*. – 2008. – No. 7. – P. 71-76.

[145] Le Rich H.H., Weir A.N. A method of studying trace elements on soil fractions // *J. Soil. Sci.* – 1963. - V. 14, No. 12. – P. 71-75.

[146] Lead in the Environment. – Moscow: *Nauka*, 1987. – 181 p.

[147] Lean D., Rahayu U., Winch S., Ridal J., Blais J., Hassan N., Murimboh J., Chakrabarti C.L. Interpretations of biological and chemical indicators for metal toxicity // *Proceedings of the 6th International Conference ICOBTE-2001 "Biogeochemistry of Trace Elements"*. – Canada, 2001. - P. 118.

[148] Lebedeva L.A., Lebedev R.N., Edemskaya N.L., Grafskaya G.A. Effect of lime materials and organic fertilizers on the content of cadmium in plants // *Agrokhimiya*. - 1997. - No. 10. - P. 45-51.

[149] Leenheer J.A. Oxidative diagenesis of metal binding structures in natural organic matter // *Proceedings of the 5th International Conference of Biogeochemistry of Trace Elements*. – Vienna, Austria, 1999. – P. 368-369.

[150] Li Z., McLaren R.G., Metherell A.K. Fractionation of cobalt and manganese in New Zealand soils // *Aust. J. Soil Res.* – 2001. - V. 39. - P. 951-967.

[151] Lindsay W.L. Chemical Equilibria in Soil. – New York, 1979. – 449 p.

[152] Lindsay W.L., Norvell W.A. Reactions of DTPA chelates of Fe, Zn, Mn, and Co with soils // *Soil Sci. Soc. Am. Proc.* - 1972. - No. 36. - P. 778-783.

[153] Loboda B.P. Use of zeolite raw materials in plant growing // *Agrokhimiya*. – 2000. - No. 6. – P. 78-91.

[154] Loganathan P., Burau R.G., Fuerstenau D.W. Influence of pH on the sorption of Co, Zn, and Ca by a hydrous manganese oxide // *Soil Sci. Am. Proc.* - 1977. - No. 41. - P. 57-62.

[155] Ma Y.B., Uren N.C. The fate and transformations of zinc added to soils // *Austr. J. Soil Res.* – 1997. - V. 35. - P. 727 – 738.

[156] Machado P.L.O. de A., Pavan M.A. Zinc sorption by some soils of Parana // *Revista Brasileira de Ciencia do Solo.* - 1987. - V. 11, No. 3. - P. 253-256.
[157] Manceau A., Lanson B., Schlegel M.L., Harge J.C., Musso M., Eybert-Berard L., Hazemann J.L., Chateigner D. Lamble G.M. Quantitative Zn speciation in smelter-contaminated soils by EXAFS spectroscopy // *Am. J. Sci.* - 2000. - V. 300. - P. 289-343.
[158] Manceau A., Marcus M.A., Tamura N. Quantitative speciation of heavy metals in soils and sediments by synchrotron X-ray techniques // *Applications of Synchrotron Radiation in Low-Temperature Geochemistry and Environmental Science: Reviews in Mineralogy and Geochemistry.* - Washington, DC, 2002. - V. 49. - P. 341-428.
[159] Manskaya R.M., Drozdova T.V., Emel'yanova T.V., Emel'yanova M.P. Copper binding by different natural organic compounds // *Pochvovedenie.* - 1958. - No. 6. - P. 41-48.
[160] Manucharov A.S., Kharitonova G.V., Chernomorenko N.I., Zemlyanukhin V.N. Effect of Adsorbed Zinc and Lead Cations on the Surface Properties of Minerals and Water Vapor Sorption // *Eur. Soil Sci.* - 2001. - No. 4. - P. 639-699.
[161] McBride M.B. Chemisorption and precipitation of Mn^{2+} at $CaCO_3$ surfaces // *Soil Sci. Soc. Am. J.* – 1979. - No. 41. – P. 693–698.
[162] McBride M.B. Chemisorption of Cd^{2+} on calcite surface // *Soil Sci. Soc. Am. J.* - 1980. - V. 44, No. 1. - P. 26-28.
[163] McBride M.B. Copper in solid and solution phases of soil // *Copper in Soils and Plants* / Ed. by Logeragan Y.F., Robson A.D., Grahm K.D. – New York: Academic Press, 1981. - P. 25-43.
[164] McBride M.B. Reactions controlling heavy metals solubility in soils // *Adv. Soil. Sci.* - 1989. - V. 10. - P. 2-47.
[165] McLaren R.G., Crawford D.V. Studies on soil copper: I. The fractionation of copper in soils // *J. Soil Sci.* - 1973. - V. 24, No. 2. - P. 172-181.
[166] Meers E., Unamuno V.R., Laing G.Du, Vangronsveld J., Vanbroekhoven K., Samson R., Diels L., Geebelen W., Ruttens A., Vandegehuchte M., Tack F.M.G. Zn in the soil solution of unpolluted and polluted soils as affected by soil characteristics // *Geoderma.* – 2006. – V. 136. - P. 107–119.
[167] Mellis E.V., Casagrande J.C., Cruz M.C.P. Iron oxides, pH and organic matter effects on nickel adsorption // *Proceedings of the 7th International*

Conference on the Biogeochemistry of Trace Elements, Uppsala, Sweden, June 15-19, 2003. - V. 3. - P. 20-21.
[168] Methodological Guidelines on the Determination of heavy Metals in Agricultural Soils and Crops. - Moscow: TsINAO, 1992. – 61 p.
[169] Microelements in the Soils of the USSR. Vol. 1: Microelements in Soils of European Soviet Union / Ed. by Kovda V.A., Zyrin N.G. - Moscow: MGU, 1973.
[170] Miller P.W., Martens D.C., Zelazny L.W. Effect of sequence in extraction of trace metals from soils // *Soil Sci. Am. J.* - 1986. - V. 50. - P. 598-601.
[171] Mineev V.G., Alekseev A.A., Trishina T.A. Zinc in the Environment // *Agrokhimiya.* - 1984. - No. 3. - P. 94-104.
[172] Mineev V.G., Kochetavkin A.V., Nguen Van Bo. Use of natural zeolites for preventing the contamination of soils and plants with heavy metals // *Agrokhimiya.* - 1989. - No. 8. - P. 89-95.
[173] Ming D.W., Mumpton F.A. Zeolites in soils // Minerals in Soil Environments. – Ed. by Dixon J.B. and Weed S.B. - SSSA BOOK Series. – Madison, WI. - P. 873-911.
[174] Minkin M.B. Physicochemical studies of the exchangeable complex of soils in the Lower Don region and some problems of their reclamation / *Doctoral Dissertation in Biology.* - 1974. - 284 p.
[175] Minkin M.B., Endovitskii A.P., Kalinichenko V.P. Calcium–Carbonate Equilibrium in Soil Solutions. - Moscow: *MSKhA,* 1995. – 212 p.
[176] Minkina T.M., Adriano D.C., Samokhin A.P., Nazarenko O.G. Organic matter interactions with zinc and lead // *Proceeding of the 10th International Meeting of the International Humic Substances Society* (IHSS 10). - Toulouse, France, July 2-28, 2000a. - P. 487-489.
[177] Minkina T.M., Kalinitchenko V.P., Nazarenko O.G., Nilityuk N.V., Samokhin A.P. Peculiarities of heavy metals mobility investigation in calcareous chernozem // *Abstracts Book of the 16th World Congress of Soil Science.* - Montpellier, France, 1998. - V. 1. - P. 120.
[178] Minkina T.M., Kryshchenko V.S., Samokhin A.P., Nazarenko O.G. Technogenic Contamination of Soil with Heavy Metals / *Textbook.* – Rostov-on-Don: Kopitsentr, 2003. – 75 p.
[179] Minkina T.M., Motuzova G.V, Nazarenko O.G., Mandzhieva S.S. Transformation of heavy metal compounds in soils of chernozem // *Proceedings of the 14th International Conference on Heavy Metals in the Environment.* - Taipei, Taiwan, 2008c. - P. 539-542.

[180] Minkina T.M., Motuzova G.V. and Nazarenko O.G. Interaction of heavy metals with the organic matter of an ordinary chernozem // *Eur. Soil Sci.* - 2006. - No. 7. - P. 702-710.
[181] Minkina T.M., Motuzova G.V., Nazarenko O.G., Kryshchenko V.S., Mandzhieva S.S. Forms of heavy metal compounds in soils of the steppe zone // *Eur. Soil Sci.* - 2008a. - V. 41. - N. 7. - P. 708–716.
[182] Minkina T.M., Samokhin A.P. and Nazarenko O.G. Influence of soil contamination by heavy metals on organic matter // *Man and Soil at the Third Millennium: Proceedings of the Third International Congress of the ESSC*, 2002. – V. 2. – Valencia, Spain, 28 March – 1 April, 2000b. – P. 1859-1865.
[183] Minkina T.M., Motuzova G.V., Nazarenko O.G., Kryshchenko V.S., Mandzhieva S.S. Combined approach for fractioning metal compounds in Soils // *Eur. Soil Sci.* - 2008b. - V. 41. – N. 11. - P. 1170–1178.
[184] Minkina T.M., Motuzova G.V., Nazarenko O.G., Samokhin A.P., Kryshchenko V.S., Mandzhieva R.S. Effect of different ameliorants on the mobility of zinc and lead in contaminated chernozem // *Agrokhimiya.* - 2007. - No. 10. - P. 67-75.
[185] Minkina T.M., Nazarenko O.G., Mandzhieva R.S. Fractional composition of heavy metals in soils contaminated with emissions from the Novocherkassk power station // *Vestn. Yuzhn. Nauchn. Tsentra, Ross. Akad. Nauk.* – 2007. - V. 3, No. 4. - P. 53-64.
[186] Minkina T.M., Pinsky D.L., Samokhin A.P., Statovoi A.A. Sorption of copper, zinc, and lead by ordinary chernozem under mono- and polyelement contamination // *Agrokhimiya.* - 2005. - No. 8. - P. 58-64.
[187] Minkina T.M., Samokhin A.P. and Nazarenko O.G. Influence of soil contamination by heavy metals on organic matter // *Man and Soil at the Third Millennium: Proceedings of the Third International Congress of the ESSC*, 2002. – V. 2. – Valencia, Spain, 28 March – 1 April, 2000b. – P. 1859-1865.
[188] Mossop K.F., Davidson C.M. Comparison of original and modified BCR sequential extraction procedures for the fractionation of copper, iron, lead, manganese and zinc in soils and sediments // *Anal. Chim. Acta.* – 2003. – V. 478. – P. 111–118.
[189] Motuzova G.V. Microelements in Soils: Systemic Organization, Ecological Significance, and Monitoring. – Moscow, 1999, P. 168.
[190] Motuzova G.V. Microelements in subtropical soils of Western Georgia / *Extended Abstract of Candidate's Dissertation in Biology.* – Moscow, 1972. – 24 p.

[191] Motuzova G.V. Principles and Methods of Soil-Chemical Monitoring. - Moscow: MGU, 1988. – 99 p.
[192] Motuzova G.V., Aptikaev R.S. Classification of soils by the microelement ratios (with arsenic as an example) // *Problems in Biogeochemistry and Geochemical Ecology.* – 2006. - No. 2. - P. 59-67
[193] Motuzova G.V., Aptikaev R.S., Karpova E.A. Fractionation of soil arsenic compounds // *Eur. Soil Sci.* - 2006. - No. 4. - P. 387-396.
[194] Motuzova G.V., Bezuglova O.S. Ecological Monitoring of Soils. - Moscow: *Gaudeamus,* 2007. - 237 p.
[195] Motuzova G.V., Degtyareva A.K. Effect of 0.1 N H_2SO_4 and Tamm and Mehra-Jackson reagents oniron compounds in soddy-alluvial soil // *Vestn. Mosk. Univ., Ser.* 17: Pochvoved. - 1991 - No. 1. - P. 36-48.
[196] Motuzova G.V., Popova A.A. Relationship between the mobility of Zn and the chemical properties of soils // *Agrokhimiya.* - No.8. - 1988. - P. 81-88
[197] Motuzova G.V., Smirnova E.V. Copper, zinc, and manganese in geochemically coupled series of some soils in the Sikhote Alin Reserve // *Geochemistry of Heavy Metals in Natural and Technogenic Landscapes.* - Moscow: MGU, 1983. - P. 37-46.
[198] Nielsen J.D. Specific zinc adsorption as related to the composition and properties of clay and silt in some Danish soils // *Acta Agric. Scand.* - 1990. - V. 40. - No. 1. - P. 3-9.
[199] Nikityuk N.V. Mobility of heavy metals in calcareous chernozemic soils and methods of its assessment / *Extended Abstract of Candidate's Dissertation in Agriculture.* - Krasnodar: KGAU, 1998. - 18 p.
[200] Nosovskaya I.I., Solov'ev G.A., Egorov V.S. Effect of the long-term regular application of different mineral fertilizers and manure on the accumulation of Pb, C, Ni, and Cr in the soil and their economic balance // *Agrokhimiya.* – 2001. - No. 1. – P. 82-91.
[201] Obukhov A.I. Resistance of chernozems to the contamination with heavy metals // Problems in the Conservation, Rational Use, and Remediation of Chernozems. - Moscow: *Nauka,* 1989. - P. 33-41.
[202] Obukhov A.I., Tsaplina M.A. Transformation of technogenic heavy metal compounds in soddy-podzolic soil // *Vestn. Mosk. Univ., Ser.* 17: Pochvoved. - 1990 - No. 3. - P. 39-44.
[203] Okorkov V.V., Mazirov M.A. Some physicochemical aspects of remediation of soils contaminated with heavy metals // *Proceedings of the 1st International Geoecological Conference "Geoecological*

Problems of Environmental Contamination with heavy Metals". – Tula: TGU, 2003. – P. 529-533.
[204] Orlov D.S. Soil Humus Acids. - Moscow: MGU, 1974.
[205] Ovcharenko M.M. Mobility of heavy methods in the soil and their availability to plants // *Agr. Nauka* – 1996 - No. 3. - P. 39-41.
[206] Pampura T.V., Pinsky D.L., Ostroumov V.G., Gershevich V.D., Bashkin V.N. Experimental study of chernozem buffering properties under contamination with copper and zinc // *Pochvovedenie.* - 1993. - No. 2. - P. 104-111.
[207] Panin M.S., Bairova A.M. Sorption of lead by chestnut soils of the Semipalatinsk Irtysh region depending on the particle-size distribution of the zeolites added // *Agrokhimiya.* - No. 10. - 2005. - P. 92-96.
[208] Panin M.S., Kushnareva A.Yu. Zinc compound forms in dark chestnut soils under mono- and polyelement contamination // *Agrokhimiya* - 2007. - No. 6. - P. 68-73.
[209] Papadopoulos P., Rowell D.L. The reaction of cadmium with calcium carbonate surfaces // *J. Soil Sci.* – 1988. - No. 39. – P. 23–36.
[210] Peive Ya.V., Rin'kis G.Ya. Field laboratory for the determination of plant-available microelements in soils // *Microelements in Plant Growing.* – Riga, 1958. – 354 p.
[211] Pendyurin E.A. Agroecological conditions of growing agricultural crops on chernozems contaminated with heavy metals / *Extended Abstract of Candidate's Dissertation in Biology.* – Belgorod, 1986. - 21 p.
[212] Perelomov L.V., Pinsky D.L. Mn, Pb, and Zn compounds in gray forest soils of the Central Russian Upland // *Eur. Soil Sci.* - 2003. - No. 6. - P. 610-618.
[213] Perez de Mora A., Madejon E., Madrid F., Cabrera F. Use of organic amendments to remediate heavy metal and toxic element contaminated soils // *Proceedings of the 7th International Conference on the Biogeochemistry of Trace Elements, Uppsala, Sweden*, June 15-19, 2003. - V. 2. - P. 226-227.
[214] Peterburgskii A.V. Laboratory Manual on Agronomical Chemistry. - Moscow: *Kolos,* 1968. – 421 p.
[215] Piccolo A., Stevenson F.J. Infrared spectra of Cu^{2+}, Pb^{2+} and Ca^{2+} complexes of soil humic substances // *Geoderma.* - 1982. - V. 27, No. 3. - P. 195-208.
[216] Pickering W.F. Zinc interactions with soil and sediment components // *Zinc in Soils* / Ed. by Nriagu J.O. – New York: Wiley, 1980. - P. 40-57.

[217] Pierzynski G.M., Baker L.R., Hettiarachchi G.M., Scheckel K.G. The use of soil amendments for the remediation of heavy metal-contaminated sites // *Proceeding of the 14th International Conference on Heavy Metals in Environment "ICHMET"*. – Taiwan, 2008. – P. 1-3.

[218] Pierzynski G.M., Thomas S.J., Vance G.F. Soils and Environmental Quality, Third Edition. Taylor&Francis, 2005. 480p.

[219] Pingitore N.E., Eastman M.P. The experimental partitioning of Ca^{2+} into calcite // *Chem. Geol.* – 1986. - No. 45. – P. 113–120.

[220] Pinsky D.L. Ion Exchange in Soils. - Pushchino: *Ross. Akad. Nauk*, 1997. - 166 p.

[221] Pinsky D.L. Tendencies and mechanisms of cation exchange in soils / Extended Abstract of Doctoral Dissertation in Biology. - Moscow: *MGU*, 1992. – 34 p.

[222] Pinsky D.L., Antalova S., Mocik A. The state of Cd, Pb, and Zn in soils and their uptake by plants // *International Conference of Soil Conservation and Environment.* - 1989. – P. 2-5.

[223] Plekhanova I.O., Klenova O.V., Kutukova Yu.D. The Effect of Sewage Sludge on the Content and Fractional Composition of Heavy Metals in Loamy-Sandy Soddy-Podzolic Soils // *Eur. Soil Sci.* - 2001. - No. 4. - P. 440-446.

[224] Pollutants in the Environment / Ed. by Motsyuk A., Pinsky D.L. – Pushchino: AN SSSR, 1988 – 195 p.

[225] Ponizovskii A.A., Dimoyanis D.D., Tsadilas K.D. The use of zeolite for the detoxification of lead-contaminated soils // *Eur. Soil Sci.* - 2003. - No. 4. - P. 439-443.

[226] Ponizovskii A.A., Mironenko E.V. Mechanisms of lead(II) sorption in soils // *Eur. Soil Sci.* - 2001. - No. 4. - P. 371-381.

[227] Ponizovskii A.A., Polybesova T.A. Seasonal changes in soil solution composition and soil particle surface properties of gray forest soil under agricultural use // *Pochvovedenie.* - 1990. - No. 12. - P. 36-45.

[228] Ponomareva V.V., Plotnikova T.A. Procedure and some results of chernozem humus fractionation // *Pochvovedenie.* – 1968. - No. 11. - P. 31-37.

[229] Popovicheva L.P. Effect of ameliorants on the lead status in contaminated soddy-podzolic soils and its input to plants / *Extended Abstract of Candidate's Dissertation in Biology.* – Moscow, 1988. – 25 p.

[230] Prokhorov V.M., Gromova E.A. Effect of pH and salt concentrations on the sorption of zinc by soils // *Pochvovedenie.* – 1971. - No. 1.

[231] Protasova N.A., Gorbunova N.S. Nickel, lead, and cadmium forms in chernozems of the Central Chernozemic region // *Agrokhimiya.* – 2006 - No. 8. - P. 68-76

[232] Provisional recommendations for the use of arable soils contaminated with heavy metals // *Materials of the Interdepartmental Scientific and Technical Conference on the Problems of Soil Contamination.* - 1990. - P. 51-62.

[233] Pryanishnikov D.N. Selected Works. – Moscow: *Sel'khozgiz,* 1953. – V. 3. – 252 p.

[234] Pueyo M., Mateu J., Rigol A., Vidal M., Lopez-Sanchez J.F., Rauret G. Use of the modified BCR three-step sequential extraction procedure for the study of trace element dynamics in contaminated soils. // *Environ. Pollut.* – 2008. – V. 152. – P. 330-341.

[235] Quy R.D., Chakrabarti C.L., Mc Bain D.C. An evaluation of extraction techniques for the fractionation of Cu and lead in model sediment systems // *Water Res.* – 1978. - V. 12. - P. 21-24.

[236] Ramamoorthy S., Rust B.R. Heavy metal exchange processes in sediment-water systems // *Environ. Geol.* - 1978. - No. 2. - P.165-172.

[237] Reeder R., Prosky J.L. Compositional sector zoning in dolomite // *J. Sediment Petrol.* – 1986. - No. 56. – P. 237–247.

[238] Reeder R.J., Grans J.C. Sector zoning on calcite cement crystals: Implications for trace element distributions in carbonates // *Geochim. Cosmochim. Acta.* – 1987. - No. 51. – P. 187–194.

[239] Reference Book on Industrial Mineral Fields in Rostov Oblast / Ed. by Pushkarskii E.N. – Rostov-on-Don, 2000. – 247 p.

[240] Report of Large-Scale Geochemical and Radiometric Studies of Ecological Situation in the City of Novocherkassk in 1991–1994: Research Report. – *Novocherkassk: Prospekt,* 1995. - V. 1–5. - P. 178.

[241] Rerikh V.I. Zinc and Cobalt Forms in Soils / *Extended Abstract of Candidate's Dissertation in Biology.* - Moscow, 1976 – 31 p.

[242] Reshetnikov R.I. Copper forms in polluted and background soddy-podzolic soils // *Biol. Nauki.* - 1990. - No. 4. – P. 114-123.

[243] Retstse K., Krystya K. Control of Soil Pollution. – Moscow: *Agropromizdat,* 1986. – P. 69-87.

[244] Rin'kis G.Ya. Methods of determining microelements in biological samples // *Microelements and the Natural Radioactivity of Soils: Proceedings of the 3rd Interuniversity Workshop.* – Rostov-on-Don, 1962. – P. 238-241.

[245] Roberts D.R., Scheinost A.C., Sparks D.L. Zn speciation in a smelter-contaminated soil profile using bulk and microscopic techniques // *Environ. Sci. Technol.* - 2002. - V. 36. – P. 1742-1750.
[246] Rukhovich O.V. Cation exchange in copper-contaminated soils: *Extended Abstract of Candidate's Dissertation in Biology.* – Moscow, 1993. – 158 p.
[247] Sadovnikova L.K. Heavy metals // *Soil-Ecological Monitoring.* - Moscow, 1994. - P. 105-120.
[248] Sadovnikova L.K. Use of soil extracts in the study of heavy metal compounds // *Khim. Sel'sk. Khoz.* - No. 2. - 1997. - P. 37-40.
[249] Samokhin A.P. Transformation of heavy metal compounds in soils of the Lower Don region / *Extended Abstract of Candidate's Dissertation in Biology.* – Rostov-on-Don, 2003. - 24 p.
[250] Santillan-Medrano J., Jurinak J.J. The chemistry of lead and cadmium in soil solid phase formation // *Soil Sci. Soc. Am. Proc.* - 1975. - V. 39. – No. 5. – P. 851-856.
[251] Schachschabel P. Die Bestimmung des Manganversorgung Grades der Boden // *Rapp. 6th International Congress of Soil Scientists.* - Paris, 1956. – P. 302.
[252] Scheinost A., Kretzschmar R. Metal speciation in contaminated soil complementary results from selective sequential extractions and XAFS // *Proceedings of the 6th International Conference ICOBTE-2001 "Biogeochemistry of Trace Elements".* – Canada, 2001. - P. 65.
[253] Scheinost A.C., Kretzchmar R.S., Pfister S. Combining selective sequential extractions, X-ray absorption spectroscopy, and principal component analysis for quantitative Zn speciation in soil // *Environ. Sci. Technol.* - 2002. - V. 36. – P. 5021-5028.
[254] Schizer M., Skinner S.I.M. Organo-metallic interactions in soils // *Soil Sci.* - 1967. - V. 103, No. 4. - P. 80-85.
[255] Schlegel ML, Manceau A., Charlet L, Chateigner D., Hazemann J.I. Sorption of metal ions on clay minerals: III. Nucleation and epitaxial growth of Zn phyllosilicate on the edges of hectorite //*Geochim. Cosmochim. Acta.* 2001. – V. 65. – P. 4155-4170.
[256] Shestakov E.I. Interaction between manganese ions and water-soluble organic matter and fulvic acids from podzolic soils / *Extended Abstract of Candidate's Dissertation.* - Moscow, 1984. – 21 p.
[257] Shestakov E.I., Karpukhin A.I., Kaurichev I.S., Rachinskii V.V. Migration and transformation of manganese compounds in podzolic soils // *Pochvovedenie.* - 1989. - No. 12. - P. 35-46.

[258] Shibaeva I.N. Comparison of two methods for extraction microelements from soil organic matter // *Vestn. Mosk. Univ., Ser. 17: Pochvoved.* – 1990. - No. 3. - P. 32-38.
[259] Shuman L.M. Fractionation method for soil microelements // *Soil Sci.* – 1980. - V. 140, N 1. - P. 11-22.
[260] Shuman L.M. Sodium hypochlorite method for extracting microelements associated with soil organic matter // *Soil. Sci. Soc. Am. J.* - 1983. - V. 47. - P. 10-17.
[261] Shuman L.M. Zinc, manganese, and copper in soil fractions // *Soil Sci.* – 1979. - V. 127 – P. 10-17.
[262] Singh M.V., Abrol I.P. Solubility and adsorption of zinc in a sodic soil // *Soil Sci.* - 1985. - V. 140, No. 6. - P.285-291.
[263] Singhal J.P., Kumar D. Thermodynamics of exchange of Zn with Mg bentonite and Mg illite // *Geoderma.* - 1977. - No. 17. - P.235-258.
[264] Sizov A.P., Khomyakov D.M. Problems in the Control of Soil and Crop Contamination. – Moscow: MGU, 1990. – 51 p.
[265] Solov'ev G.A. Use of complex extracts for determining available microelements in soils // *Monitoring of the Background Contamination of Natural Environment.* - Leningrad: Gidrometeoizdat, 1989. – No. 56. - P. 216-227.
[266] Soon Y.K., Bates T.E. Chemical pools of cadmium, nickel, and zinc in polluted soils and some preliminary indications of their availability to plans. // *J. Soil Sci.* - 1982. - V. 33. - P. 477-488.
[267] Sparks D.L. Environmental soil chemistry. - Elsevier: Academic press, 2003. - 352 p.
[268] Sposito G. The Chemistry of Soil. – New York: Oxford Univ. Press, 1989. - 338 p.
[269] Sposito G. The Surface Chemistry of Soils. – New York: Oxford Univ. press, 1984. – 320 p.
[270] State Report "On the State of the Natural Environment in the City of Novocherkassk in 1997". - Novocherkassk, 1998. – 27 p.
[271] Steinnes E., Njastad O. Enrichment of metals in the organic surface layer of natural soil: Identification of contribution from different sources // *Analyst.* – 1995. - Vol. 120.
[272] Stevenson F.J. Stability constants of Pb and Cd complexes with humic acids // *Soil Sci. Soc. Am. J.* - 1976. - V. 40. - P. 665-672.
[273] Stipp S.L., Hochella M.F.Jr., Parks G.A., Leckie J.O. Cd^{2+} uptake by calcite, solid-state diffusion, and the formation of solid-solution: Interface processes observed with near-surface sensitive techniques

(XPS, LEED, and AES). *Geochim. Cosmochim. Acta.* – 1992. No. 56. – P. 1941–1954.
[274] Strand V., Zolotareva B.N., Lisovskii A.E. Effect of water-soluble lead, cadmium, and copper salts on their input to plants and the yield of crops // *Agrokhimiya.* - 1991. - No. 4. - P. 76-83.
[275] Suslina L.G., Anisimova L.N., Kruglov R.V., Anisimov V.S. Accumulation of Cu, Zn, Cd, and Pb by barley from soddy-podzolic and peat soils under potassium application at different pH // *Agrokhimiya.* - 2006. - No. 6. - P. 69-79.
[276] Tan K.H., King L.D., Morris H.D. Complex reactions of Zn with organic matter extracted from sewage sludge // *Soil Sci. Am. Proc.* - 1971. - V.35, No. 5. - P. 631-639.
[277] Tarasevich Yu.I. Structure and Chemistry of Surface Layered Silicates. - Kiev: *Naukova Dumka*, 1988. – 247 p.
[278] Tarasevich Yu.I., Ovcharenko F.D. Adsorption on Clay Minerals. - Kiev: *Naukova Dumka*, 1975. – 375 p.
[279] Tarr M.A., Lindsey M.E., Xu G. Sorption of hydrophobic pollutants to natural organic matter: effects on pollutant degradation // *Proceeding of the 10th International Meeting of the International Humic Substances Society (IHSS 10).* - Toulouse, France, July 2-28, 2000. – P. 297-300.
[280] Tessier A., Campbell P.G.C., Bisson M. Sequential extraction procedure for the speciation of particulate trace metals // *Anal. Chem.* – 1979. - V. 51, N. 7. - P. 844-850.
[281] Tessier A., Rapin F., Carignan R. Trace elements in oxic lake sediments: possible adsorption onto iron oxyhydroxides // *Geochim. Cosmochim. Acta.* - 1985. - V.49. - P. 183-194.
[282] Theory and Practice of Chemical Soil Analysis / Ed. by Vorob'eva L.A. – Moscow: Geos, 2006. - 400 p.
[283] Toporskaya L.E., Danilova L.E., Danilova G.N. Ecological State of the Environment in Novocherkassk // *Problems in Geology and Geoecology of the Southern Russia and Caucasus: proceedings of the International Conference.* - Novocherkassk: NA-BLA, 1997. - V. 2. - P. 75-78.
[284] Tyurin I.V. Soil Organic Matter and Its Role in Soil Formation and Fertility: Theory of Soil Humus. - Moscow, 1937. - 288 p.
[285] Unguryan V.G., Kholmitskii A.M., Problems in the Cultivation and fertility Reproduction of Moldavian Soils. - *Chisinau,* 1978.
[286] Val'kov V.F. Genesis of Northern Caucasian Soils. – Rostov-on-Don: RGU, 1977. – 160 p.

[287] Val'kov V.F. Soil Science (Soils of the Northern Caucasus). – Krasnodar: *Sov. Kuban'*, 2002. – 728 p.
[288] Van Dijk. Cation binding by humic acids // *Geoderma.* - 1971. - No. 5. - P. 53-67.
[289] Van N.T.Kh. Chemical and physical properties of red ferralitic soils in the rubber plantations of the Southern Viet Nam and Cambodia. – Candidate's Dissertation in Biology. - Moscow, 1995. - 176 p.
[290] Vernadsky V.I. Sketches on Geochemistry. – Moscow, 1934. – 346 p.
[291] Violante, A., Huang P.M., Bollag J.-M. and Gianfreda L. (Eds.). Soil Mineral-Organic Matter-Microorganism Interactions and Ecosystem Health. Dynamics, Mobility and Transformation of Pollutants and Nutrients // *Develop. Soil Sci. Vol. 28A. Elsevier, Amsterdam.* 2002. 459 p.
[292] Vincler P., Lacatov B., Mady Gy., Meisel J., Mohos B. Infrared and EPR spectra of peats, peat humic acids, and metal humates // *Humus Planta.* - 1971. - No. 5. - P. 301-304.
[293] Vinogradov A.P. Geochemistry of Rare and Dispersed Chemical Elements in Soils. - Moscow, 1957. - 68 p.
[294] Vodyanitskii Yu.N. Study of Heavy Metals in Soils. - Moscow: *Pochv. Inst. im. V.V. Dokuchaeva*, 2005a. – 110 p.
[295] Vodyaniskii Yu.N. Manganese Oxides in Soils. - Moscow: *Pochv. Inst. im. V.V. Dokuchaeva*, 2005b. - 168 p.
[296] Vodyaniskii Yu.N. Role of soil components in the fixation of technogenic As, Zn, and Pb in soils // *Agrokhimiya.* – 2008. – No. 1. – P. 83-91.
[297] Vodyaniskii Yu.N., Dobrovol'skii V.V. Irn Minerals in Soils. - Moscow: *Pochv. Inst. im. V.V. Dokuchaeva*, 1998. - 216 p.
[298] Vorob'eva L.F., Rudakova T.A. Possibility of predicting the status of some chemical elements in natural waters and solutions from solubility diagrams // *Vestn. Mosk. Univ., Ser. 17: Pochvoved.,* – 1981. - No. 4. - P. 3-12.
[299] Whalley C., Grant A. Assessment of the phase selectivity of the European Community Bureau of Reference (BCR) sequential extraction procedure for metals in sediment // *Anal. Chem. Acta.* - 1994. - V. 61. - P. 2211-2221.
[300] Wilde S.A., Voigt G.K. Analysis of Soils and Plants for Foresters and Horticulturists. – Michigan: Edwards publisher, 1955. - 117 p.

[301] Wong J.W.C., Li K.L., Zhou L.X., Selvam A. The sorption of Cd and Zn by different soils in the presence of dissolved organic matter from sludge // *Geoderma.* – 2007. – V. 137. - P. 310–317.
[302] Yang J.E, Ok Y.S., Kim W.S.., Lee J.S. Remediation of heavy metal-contaminated soil amendments and management method solutions // *Proceedings of the 14th International Conference on Heavy Metals in Environment "ICHMET".* – Taiwan, 2008. – P. 166-169.
[303] Yin Y., You S., Allen H.E. Lability of heavy metals in soils: role of soil properties // *Proceedings of the 5th International Conference of Biogeochemistry of Trace Elements.* – Vienna, Austria, 1999. – P. 358-359.
[304] Zachara J.M., Kittrick J.A, Harsh J.B. The mechanism of Zn^{2+} adsorption on calcite // *Geochim. Cosmochim. Acta.* – 1988. - No. 52. – P. 2281–2291.
[305] Zakrutkin V.E., Ryshkov M.V. Ecological zoning of the Rostov oblast // *Izv. Vyssh. Uchebn. Zaved., Estestv. Nauki.* - 1997. - No. 4. - P. 83-89.
[306] Zakrutkin V.E., Shishkina D.Yu. Some aspects of copper and zinc distribution in soils and plants of agrolandscapes in Rostov oblast // *Proceedings of the International Workshop "Heavy Metals in the Environment".* – Pushchino, 1997. – P. 101-109.
[307] Zakrutkin V.E., Shkafenko R.P. Some aspects of lead distribution in soils and plants of agrolandscapes in Rostov oblast // *Proceedings of the International Workshop "Heavy Metals in the Environment".* - Pushchino, 1997. - P. 110–117.
[308] Zemberyova M., Bartekova J., Hagarov I. The utilization of modified BCR three-step sequential extraction procedure for the fractionation of Cd, Cr, Cu, Ni, Pb, and Zn in soil reference materials of different origins // *Talanta.* – 2006. – V. 70. – P. 973–978.
[309] Zhideeva V.A., Vasenev I.I., Shcherbakov A.P. Fractionation of Pb, Cd, Ni, and Zn compounds in meadow-chernozemic soils polluted by emissions of the accumulator plant // *Eur. Soil Sci.* – 2002. – No. 6. – P. 643-650.
[310] Zien H., Brunner G.W. Ermittlung der Mobilität und Bindungsformen von Schwermetallen in Böden mittels sequentieller Extraktionen // *Mitt. Dtsch. Bodenkundl. Gesellsch.* – 1991. – V. 66, No. 1. – P. 439.
[311] Zonn R.V. Iron in Soils - Moscow: *Nauka,* 1982. - 206 p.
[312] Zonn, R.V., Travlev A.P. Aluminum, Its Role in Pedogenesis, and Effect on Plants. - Dnepropetrovsk, 1992. - 224 p.

[313] Zunio H., Peirrano P., Aguilera M., Schalsha E.B. Measurement of metal complexing ability of polyfunctional molecules: a discussion of the relationship between the metal-complexing properties of extracted organic matter and soil genesis and plant nutrition // *Soil Sci.* - 1975. - No. 119. - P.210-216.

[314] Zyrin N.G. Key problems in the theory of trace elements in soil science: Report of the Doctoral Dissertation in Biology. - Moscow: MGU, 1968. – 38 p.

[315] Zyrin N.G., Chebotacheva N.A. Copper, zinc, and lead forms in soils and their availability to plants // *Content and Forms of Microelements in Soils.* - Moscow: Nauka, 1983. - P. 93-114.

[316] Zyrin N.G., Motuzova G.V., Siminov V.D., Obukhov A.I. Microelements (boron, manganese, copper, zinc) in soils of the Western Georgia // *Content and Forms of Microelements in Soils.* - Moscow: MGU, 1979. - P. 3-159.

[317] Zyrin N.G., Obukhov A.I., Motuzova G.V. Forms of microelement compounds in soils and methods of their study // *Proceedings of the X International Congress of Soil Science.* - Moscow, 1974. - V. 2. - P. 48-49.

[318] Zyrin N.G., Serdyukova A.V. Sokolova T.A. Lead sorption and the state of adsorbed element in soils and soil components // *Pochvovedenie.* - 1986. - No. 4. - P. 39-44.

INDEX

A

absorption spectroscopy, xiii, 116, 170
acceptor, 27, 31
accounting, 38
acetate, 10, 38, 39, 40, 41, 48, 60, 80
acid, 3, 5, 8, 11, 12, 13, 14, 18, 20, 30, 31, 32, 34, 39, 40, 41, 49, 51, 60, 73, 125, 128, 138, 153, 155, 158
activation, 18, 28
activation energy, 28
active centers, 137
adsorption, 3, 5, 6, 13, 15, 16, 17, 18, 20, 28, 30, 78, 114, 117, 121, 127, 129, 146, 150, 153, 155, 157, 159, 163, 166, 171, 172, 174
aerosol, 102, 106
aerosols, 84
agent, 40, 106, 126
agents, 28, 37, 43, 78, 102, 126
aging, 17
agricultural, xiv, 9, 58, 125, 167, 168
agricultural chemistry, 9
agricultural crop, 167
agriculture, 54
agrochemicals, 155
air, 4, 15, 84, 114, 125
alkali, 14, 22, 40
alkaline, 3, 8, 13, 15, 16, 19, 37, 56, 76, 128, 130, 155
alkalinity, 73
alluvial, 12, 54, 87, 88, 101, 102, 103, 104, 107, 111, 150, 166
ALS, 31
aluminosilicate, 139
aluminosilicates, 33, 34
aluminum, 16, 17, 18, 29, 30, 46, 153, 174
amendments, 153, 160, 167, 168, 174
amino groups, 14
ammonium, 10, 33, 39, 41, 48, 127
amorphous, 4, 15, 27, 40, 45
anthropogenic, xiv, 53, 54, 109, 125, 149
apatite, 160
aqueous solutions, 157, 160
aragonite, 160
argument, 11
arid, 8, 19, 20, 54, 156
aromatic rings, 14, 75
arsenic, 22, 153, 166
ascorbic, 34
ascorbic acid, 34
ash, 140
assessment, xi, xv, 25, 87, 108, 113, 121, 149, 156, 166
atmosphere, 160
atoms, 1, 6, 7, 14, 18, 67

Index

availability, xi, xiii, xiv, 2, 9, 10, 17, 26, 28, 40, 153, 157, 167, 171, 175

B

barium, 160
barley, 60, 80, 152, 172
barrier, xiii, xiv, 78, 125, 126
behavior, xii, xiv, xv, 1, 2, 20, 108, 146
binding, xii, xiii, xiv, xv, 3, 9, 13, 14, 26, 29, 30, 31, 37, 41, 46, 49, 73, 101, 106, 108, 115, 121, 124, 126, 137, 140, 147, 162, 163, 173
bioavailability, 153
biosphere, ix, xi
biota, 4
biotic, ix, xi
blocks, 137
bonds, xi, xii, xiii, 4, 17, 18, 30
buffer, 10, 37, 38, 39, 41, 48

C

Ca^{2+}, 19, 55, 56, 79, 88, 137, 167, 168
cadmium, 48, 151, 152, 154, 158, 162, 167, 169, 170, 171, 172
calcium, 72, 128, 129, 141, 152, 158, 167
calcium carbonate, 167
carbon, 30
carboxyl, 128
carrier, 4, 31
cation, 3, 14, 17, 37, 127, 128, 151, 158, 168
cattle, 139, 140
Caucasian, 22, 172
Caucasus, 157, 172, 173
CEC, 3, 15, 17, 87, 88, 121, 127
cement, 169
certificate, 156
CH3COOH, 33, 34, 44
changing environment, 3
charged particle, 31
chelates, 2, 14, 47, 51, 75, 162
chelating agents, 154

chemical interaction, 25
chemical properties, 154, 166
chemisorption, 13, 16, 18, 28, 49, 71, 137, 141, 146, 147
chloride, 8, 160
classes, 7
classical, 128
classification, 26, 31, 49, 108, 157
clay, 3, 4, 15, 16, 17, 18, 21, 33, 45, 54, 58, 59, 78, 87, 88, 89, 101, 107, 111, 127, 129, 138, 150, 156, 158, 166, 170
Co, 10, 19, 38, 49, 140, 156, 162
CO2, 140
coal, xiv, 126, 153
cobalt, 151, 162
combustion, 160
competition, 73, 129
competitor, 78
compliance, 31
composting, 60
concentrates, 20
concentration, xiv, 3, 5, 7, 8, 9, 10, 11, 18, 25, 28, 41, 48, 71, 72, 114, 115, 116, 137
conservation, 73
contaminant, 126
contaminants, xv
control, 25, 80, 87, 102, 105
correlation, 3, 13, 40
correlations, 3, 10
Coulomb interaction, 6
covalent, 4, 6, 13, 14, 27
covalent bond, 6, 13, 14
covering, 15
CRC, 159
crops, xv, 154, 167, 172
crystal growth, 6
crystal lattice, 15, 27, 45, 67, 127
crystalline, 15
crystallization, 15, 157
crystals, 4, 137, 169
cycles, 155

D

damping, 153

decomposition, 19, 35, 45, 141
defects, 15
deficit, 25
degradation, 102, 172
deposits, 88, 101, 128
derivatives, 13, 152
desorption, 10, 28, 154
destruction, 74
detoxification, 126, 168
diagenesis, 162
differentiation, 46
diffusion, 30, 31, 171
digestion, 18, 35
dipole, 27
dispersion, 29, 49
displacement, 15, 37, 41, 43, 74
distortions, 18
distribution, 2, 3, 20, 28, 59, 101, 103, 118, 152, 154, 167, 174
diversity, ix, xi, 7, 46, 48, 138
donor, 27, 31
draft, 11
drying, 129
duration, 111

E

earth, 14
ecological, ix, xi, xii, xiii, xv, 4, 7, 9, 11, 24, 25, 26, 28, 38, 40, 51, 53, 56, 78, 84, 94, 107, 108, 124, 149, 159
ecosystem, xiii, xv, 53, 124, 125
electron, 2, 13, 31
electron microscopy, 31
electronegativity, 13, 71
electrons, 31
electrostatic force, 4, 6, 45
emission, 84, 94, 106
emission source, 94
energy, 4, 17, 28, 31, 37
enterprise, 84
environment, ix, xi, xiv, 114, 153, 161
environmental conditions, xiv, 2
environmental control, 155
environmental factors, 3

epitaxial growth, 170
EPR, xiii, 14, 18, 173
equilibrium, xiii, 1, 7, 28, 30, 46, 49, 53, 72, 77, 82, 109, 114, 116, 121, 129, 150, 158
ESR, 16, 153
estimating, 51
evaporation, 33, 44
evolution, 147
EXAFS, xiii, 6, 31, 50, 155, 163
exchange rate, 17
experimental condition, 82, 146
experimental design, 59, 139
exposure, 111
extraction, xi, xii, xiii, 10, 15, 16, 31, 37, 38, 39, 40, 41, 42, 43, 44, 45, 46, 48, 50, 51, 64, 65, 66, 146, 151, 156, 157, 164, 165, 169, 171, 172, 173, 174

F

facies, 129
feeding, 60
ferrite, 19
fertility, 9, 30, 102, 151, 172
fertilizer, 140
fertilizers, 127, 128, 162, 166
films, 15
fixation, xiv, xv, 6, 13, 15, 18, 30, 51, 58, 67, 71, 75, 77, 78, 104, 105, 106, 107, 126, 130, 138, 140, 141, 146, 147, 149, 150, 158, 173
fractional composition, xiv, 20, 29, 43, 71, 94, 104, 105, 130, 161
fractionation, xii, xv, 2, 10, 25, 26, 30, 31, 32, 37, 38, 43, 44, 45, 46, 47, 50, 51, 57, 67, 107, 108, 109, 115, 116, 123, 130, 149, 163, 165, 168, 169, 174
Fulvic acid, 13
fusion, 18, 35

G

gases, 94
gastrointestinal, 153

gel, 16
geochemical, ix, xi, 1, 4, 22, 25, 156
geochemistry, 1, 9, 157
geology, 9
goals, 37
granules, 128
groundwater, 54
groups, ix, xi, xii, xiv, 1, 6, 13, 14, 15, 18, 20, 26, 28, 31, 37, 43, 44, 51, 58, 70, 71, 72, 75, 94, 99, 108, 118, 120, 121, 128, 130, 133, 136, 138
growth, 6, 170
guidelines, 10

H

H_2, 10, 33, 35, 37, 40, 166
hardness, 42
harvesting, 60
heating, 33, 35, 44
heterogeneous, xiii, 25, 30, 53, 114, 121
horizon, 4, 80
humate, 35
humic acid, 13, 14, 74, 76, 105, 126, 146, 153, 155, 158, 171, 173
humic substances, 29, 32, 77, 152, 167
humus, 2, 13, 17, 23, 29, 54, 73, 78, 87, 88, 89, 101, 126, 138, 152, 168
hydration, 17
hydrochloric acid, 11
hydrogen, 13, 15, 27, 39
hydrogen peroxide, 15
hydrolysis, 16, 39, 60
hydrolyzed, 39
hydrophobic, 172
hydroxide, 15, 16, 19, 45, 76
hydroxides, 3, 4, 13, 15, 16, 19, 29, 33, 34, 37, 41, 45, 49, 50, 71, 73, 76, 77, 103, 105, 115, 117, 138
hydroxyl, 18, 128
hydroxyl groups, 128

I

identification, 31, 50, 108, 113
immobilization, 106, 109, 146, 157, 160
impurities, 3, 127, 160
inclusion, 19, 40, 115
indication, 30
indicators, 115, 162
indices, 38
industrial, xiv, 84, 125, 150
industry, xiv
inert, 156
inherited, 58
initial state, 77
inorganic, 4, 6, 7, 8, 9
interaction, ix, xi, 3, 5, 6, 13, 14, 15, 17, 18, 31, 44, 59, 114, 138, 140, 146, 157, 160
interactions, 6, 25, 153, 156, 160, 164, 167, 170
interface, 129
intermolecular, 14
interval, 21
ion-exchange, 5, 27
ionic, 13, 14, 19, 37, 45, 77, 115, 152
ionization, 138
iron, 15, 16, 17, 19, 20, 23, 29, 30, 31, 40, 46, 71, 76, 77, 104, 109, 153, 159, 165, 172
irradiation, 16, 33, 34
isotherms, 114, 115, 116, 121, 123

K

kaolinite, 17, 18
kinetics, 17

L

lakes, 153
landscapes, 2, 8, 11, 46, 94
leaching, 73
LEED, 172
ligands, 2, 4, 6, 7, 8, 9, 146

Index

limitation, 126
limitations, 32
linear, 114, 116
liquid phase, xiii, 4, 8, 9, 53
logical reasoning, 1
long period, 58
low molecular weight, 9

M

magnesium, 128, 152, 157, 160
magnetite, 35
management, 80, 174
manganese, 19, 29, 31, 46, 71, 76, 77, 109, 153, 154, 157, 158, 160, 162, 165, 166, 170, 171, 173, 175
mantle, 23
manure, 126, 139, 140, 141, 142, 143, 144, 145, 146, 147, 166
matrix, 29, 31, 39, 44, 45
maximum sorption, 114, 115
MCL, 59, 82, 105, 106, 140
MCLs, 11, 73, 87, 90, 91
media, xiii
metabolism, 60
metal content, 40, 42, 43, 46, 48, 54, 56, 59, 60, 67, 72, 75, 82, 105, 108, 137, 139
metal ions, 6, 8, 14, 16, 18, 20, 31, 37, 40, 41, 45, 48, 49, 50, 58, 71, 72, 74, 75, 76, 77, 82, 114, 115, 129, 170
metal oxide, 39, 49
metal oxides, 39, 49
metal salts, 2, 4, 27, 59, 127
metal-organic compounds, 141
Mg^{2+}, 19, 55, 79, 88
microorganisms, 5
microscopy, 31
migration, xi, xiii, xiv, 1, 9, 11, 25, 38, 54, 102, 129, 147, 160
millennium, 165
mine soil, 154
mine tailings, 153
mineralization, 23, 141
mineralogy, 5, 58, 139
mining, xiv
missions, 102, 106
MLC, 80
moisture, 60, 129
moisture capacity, 60
mold, 129
molecular weight, 9
molecules, 14, 18, 31, 127, 175
montmorillonite, 17, 18, 35, 58, 153, 155
morphological, 4, 128
mycelium, 129

N

Na^+, 55
natural, 2, 4, 8, 9, 10, 11, 19, 25, 28, 29, 45, 46, 54, 58, 60, 77, 109, 124, 125, 126, 127, 139, 149, 152, 155, 157, 160, 162, 163, 164, 171, 172, 173
natural environment, 2
neutralization, 41
Ni, 10, 19, 49, 139, 140, 152, 155, 156, 166, 174
nickel, 22, 152, 163, 171
nitrates, 8
nitrogen, 15
nitrogen dioxide, 15
NMR, xiii
N-N, 55, 79
nodules, 128, 129
normal, 19
norms, 159
nutrient, 157
nutrients, 10
nutrition, 127, 175

O

oat, 156
observations, 80, 84, 108, 149
occlusion, 5, 15, 16, 19, 27, 28, 30
oil, xiii, 4, 19, 25, 53, 129, 157
oils, xiv, 19, 22, 31, 54, 60, 128, 172
organ, 3, 9
organic compounds, 3, 8, 9, 15, 27, 163

organism, 139
organometallic, 3, 9
oxalate, 33
oxidation, 15, 51
oxide, 158, 162
oxygen, 6, 16, 17, 18
oxyhydroxides, 172

P

parameter, 26, 80, 110, 121
particles, xi, 3, 6, 7, 8, 10, 13, 15, 25, 26, 28, 31, 49, 71, 141, 160
PbS, 19
peat, 13, 172, 173
pH, 2, 3, 4, 8, 9, 10, 13, 14, 15, 18, 19, 20, 28, 33, 34, 38, 39, 40, 41, 44, 45, 47, 49, 50, 55, 79, 88, 105, 128, 129, 137, 138, 140, 141, 162, 163, 168, 172
pH values, 14, 16, 45
phenolic, 13
phosphate, 8, 20
phosphates, 4, 8, 19, 20, 45, 157
phosphorus, 29, 158
physical properties, 55, 79, 88, 173
physicochemical, 2, 3, 166
physiological, 28, 156
phytotoxicity, 153, 154
plants, xi, xii, xiii, xiv, xv, 3, 7, 9, 10, 11, 25, 26, 28, 40, 56, 59, 80, 125, 126, 127, 152, 154, 155, 159, 160, 162, 164, 167, 168, 172, 174, 175
play, 7, 8, 15, 20, 106, 109, 146
polarizability, 17
pollutant, 84, 172
pollutants, xiii, 2, 7, 11, 94, 102, 108, 111, 125, 146, 172
pollution, ix, xi, xiii, xiv, 2, 11, 84, 125
polyethylene, 59
polymer, 16
polymerization, 2
polynuclear complexes, 16
pools, 28, 171
potassium, 15, 127, 128, 139, 172

power, xiv, 84, 85, 86, 94, 102, 105, 106, 107, 165
power stations, 94
precipitation, 7, 16, 19, 20, 26, 28, 31, 48, 54, 126, 129, 137, 146, 154, 156, 163
prediction, xi, 11
press, 171
pressure, xiii, xiv, 35
principal component analysis, 170
probability, 115, 149
producers, xiv
production, 107
protection, 9, 124, 125, 126
protons, 14, 15, 37, 41, 129
purification, 15, 127, 157
pyrophosphate, 15

Q

quartz, 139

R

radiation, 31, 32, 163
radius, 17, 19, 37, 106, 152
range, 6, 15, 60, 87, 114, 128
ratings, 94
raw materials, 162
reaction rate, 28
reagent, 15, 16, 33, 34, 39, 40, 49, 50, 137
reagents, 2, 9, 10, 31, 32, 37, 38, 39, 40, 41, 45, 46, 49, 146, 166
reasoning, 1
reclamation, 126, 164
recrystallization, 48
redistribution, ix, xi, 4, 46, 125, 129
redox, 2, 5
regional, xiv, xv, 22, 56, 130, 138
regular, 23, 102, 166
reinforcement, 141
relationship, 114, 175
relationships, 1, 3, 25
remediation, xi, xiv, 125, 139, 146, 153, 154, 166, 168

Index

reserves, 128
residential, 84, 85
residues, 139
resins, 128
resistance, 67, 125
resources, 127
retention, 6, 23, 25, 26, 28, 32, 38, 56, 76, 77, 105, 109, 114, 115, 117, 126, 149
room temperature, 44

S

saline, 37, 49
salt, 10, 13, 27, 41, 45, 168
salts, 4, 13, 28, 37, 59, 80, 127, 147, 172
sample, 31, 35, 50
sampling, 80, 84
sand, 59
sandstones, 127
search, 3
seawater, 157
sediment, 167, 169, 173
sediments, 22, 45, 153, 154, 156, 161, 163, 165, 172
seeds, 60
selectivity, 31, 32, 126, 173
semiarid, 19
separation, xii, 15, 18, 26, 30, 37, 45, 46, 50, 51, 124, 154
series, 17, 42, 43, 77, 87, 102, 103, 107, 141, 149, 150, 166
sewage, 151, 153, 156, 172
shares, 49, 94, 104
silica, 20, 22
silicate, xii, 16, 24, 26, 35, 45, 58, 77, 78, 107, 109, 116, 121, 139, 150
silicates, 18, 22, 47, 57, 68, 69, 70, 74, 76, 95, 96, 97, 98, 99, 100, 103, 104, 108, 109, 116, 118, 119, 120, 121, 123, 131, 132, 133, 134, 135, 136
silicon, 17, 29
similarity, 73, 108
SiO2, 140
sites, 15, 28, 78, 84, 129, 150, 168
sludge, 151, 153, 156, 172, 174

sodium hydroxide, 15
soil analysis, 157
soil particles, xi, 3, 6, 7, 10, 25, 26, 28
solid phase, xiii, 2, 3, 4, 5, 6, 9, 16, 19, 26, 46, 53, 73, 114, 116, 129, 138, 141, 146, 170
solid-state, 171
solubility, 8, 9, 11, 17, 19, 20, 25, 29, 45, 49, 59, 152, 158, 163, 173
solvent, 10
solvents, xii, 10
sorbents, 19, 126, 127, 146, 155
sorption isotherms, 114, 116, 121, 123
sorption process, 155
spatial, 2, 78
speciation, 154, 160, 162, 163, 170, 172
species, 152
specific surface, 3, 58
specificity, xiv, xv, 128
spectroscopy, xiii, 31, 50, 116, 159, 162, 163
spectrum, 7
spheres, ix, xi, 49
stability, 4, 13, 14, 43, 74, 126
stabilization, 73, 141, 146, 158
stages, 30, 115
steric, 17
strategies, xi
strength, xi, xii, xiii, 6, 13, 14, 25, 26, 28, 38, 46, 49, 73, 106, 108, 114, 115, 121, 124, 140, 147
structural defect, 6
structural defects, 6
substitutes, 7
substitution, 5, 17, 18, 19, 27, 28, 30, 31, 71
subtraction, 30
suburbs, 102
sulfate, 54
superposition, 31, 39
supply, 10, 38, 56
surface layer, 171
surface properties, 168
surface water, 20, 102
suspensions, 48, 115
synchrotron, 31, 32, 35, 163

synchrotron radiation, 32

T

temperature, 44
temporal, 78
thermodynamic, 121
tolerance, xv, 149
toxic, 167
toxicity, 127, 162
trace elements, 153, 156, 162, 175
transformation, xi, xiii, xiv, xv, 4, 19, 22, 23, 28, 30, 39, 53, 60, 67, 72, 73, 77, 78, 82, 108, 109, 111, 113, 121, 129, 146, 147, 149, 150, 158, 159, 170
transformation processes, 28
transformations, xii, 23, 77, 153, 162
transition, 1, 13, 28, 82, 141, 150
transition metal, 13
translocation, 28, 59
trophic chains, 11
tundra, 23

U

Urals, 21, 23
uranium, 161
UV, 16, 33, 34
UV irradiation, 16, 33, 34

V

vacancies, 7
valence, 15
validity, 107

values, 14, 16, 45, 48, 54, 64, 65, 66, 67, 83, 93, 94, 114, 115, 121, 147
vapor, 15
variability, xiv, 2
variation, 78, 80, 129
vegetables, 154
vegetation, 59, 60
vermiculite, 58
village, 85
voids, 17

W

waste water, 157
water, 8, 9, 14, 18, 25, 28, 30, 41, 48, 127, 137, 140, 141, 157, 169, 170, 172
water-soluble, 9, 28, 30, 41, 48, 141, 170, 172
weathering, 19, 22, 23, 58, 127
wind, 84, 85, 87, 106
windows, 127
winter, 129

X

XPS, 172
X-ray absorption, xiii, 162, 170

Y

yield, 156, 172

Z

zeolites, 126, 127, 152, 157, 160, 164, 167